# Geometric Gems:
## An Appreciation for Geometric Curiosities
### Volume III: The Wonders of Circles

# Problem Solving in Mathematics and Beyond

Print ISSN: 2591-7234
Online ISSN: 2591-7242

**Series Editor:** Dr. Alfred S. Posamentier
Distinguished Lecturer
New York City College of Technology - City University of New York

There are countless applications that would be considered problem solving in mathematics and beyond. One could even argue that most of mathematics in one way or another involves solving problems. However, this series is intended to be of interest to the general audience with the sole purpose of demonstrating the power and beauty of mathematics through clever problem-solving experiences.

Each of the books will be aimed at the general audience, which implies that the writing level will be such that it will not engulfed in technical language — rather the language will be simple everyday language so that the focus can remain on the content and not be distracted by unnecessarily sophiscated language. Again, the primary purpose of this series is to approach the topic of mathematics problem-solving in a most appealing and attractive way in order to win more of the general public to appreciate his most important subject rather than to fear it. At the same time we expect that professionals in the scientific community will also find these books attractive, as they will provide many entertaining surprises for the unsuspecting reader.

*Published*

For the complete list of volumes in this series, please visit www.worldscientific.com/series/psmb

**Problem Solving in**
**Mathematics and Beyond** | Volume **36**

# Geometric Gems:
# An Appreciation for
# Geometric Curiosities
## Volume III: The Wonders of Circles

Alfred S. Posamentier
The City University of New York, USA

Robert Geretschläger
University of Graz, Austria

**World Scientific**

NEW JERSEY · LONDON · SINGAPORE · GENEVA · BEIJING · SHANGHAI · TAIPEI · CHENNAI

*Published by*

World Scientific Publishing Co. Pte. Ltd.

5 Toh Tuck Link, Singapore 596224

*USA office:* 27 Warren Street, Suite 401-402, Hackensack, NJ 07601

*UK office:* 57 Shelton Street, Covent Garden, London WC2H 9HE

**Library of Congress Cataloging-in-Publication Data**
Names: Posamentier, Alfred S., author. | Geretschläger, Robert, author.
Title: Geometric gems : an appreciation for geometric curiosities /
    Alfred S. Posamentier, Robert Geretschläger.
Description: New Jersey : World Scientific, [2024]- |
    Series: Problem solving in mathematics and beyond, 2591-7234 ; volume 32 |
    Includes index. | Contents: Volume 1. The wonders of triangles --
Identifiers: LCCN 2023043230 | ISBN 9789811279584 (vol. 1 ; hardcover) |
    ISBN 9789811281914 (vol. 1 ; paperback) | ISBN 9789811279591
    (vol. 1 ; ebook for institutions) | ISBN 9789811279607 (vol. 1 ; ebook for individuals)
Subjects: LCSH: Geometry--Popular works. | Geometry, Plane--Popular works. |
    Triangle--Popular works.
Classification: LCC QA445 .P668 2024 | DDC 516--dc23/eng/20231023
LC record available at https://lccn.loc.gov/2023043230

**British Library Cataloguing-in-Publication Data**
A catalogue record for this book is available from the British Library.

**Problem Solving in Mathematics and Beyond — Vol. 36**
**GEOMETRIC GEMS: AN APPRECIATION FOR GEOMETRIC CURIOSITIES**
**Volume III: The Wonders of Circles**
ISBN 978-981-12-9413-6 (hardcover)
ISBN 978-981-12-9476-1 (paperback)
ISBN 978-981-12-9414-3 (ebook for institutions)
ISBN 978-981-12-9415-0 (ebook for individuals)

For any available supplementary material, please visit
https://www.worldscientific.com/worldscibooks/10.1142/13866#t=suppl

Desk Editors: Balasubramanian Shanmugam/Rosie Williamson

Typeset by Stallion Press
Email: enquiries@stallionpress.com

# About the Authors

**Alfred S. Posamentier** is currently Distinguished Lecturer at the New York City College of Technology of the City University of New York. Prior to that, he was Executive Director for Internationalization and Funded Programs at Long Island University, New York. This was preceded by five years as Dean of the School of Education and Professor of Mathematics Education at Mercy University, New York. Before that, he was for 40 years at The City College of the City University of New York, where he is now Professor Emeritus of Mathematics Education and Dean Emeritus of the School of Education. He is the author and co-author of more than 80 mathematics books for teachers, secondary and elementary school students, as well as the general readership. Dr. Posamentier is also a frequent commentator in newspapers and journals on topics related to education.

After completing his BA degree in mathematics at Hunter College of the City University of New York, he took a position as a teacher of mathematics at Theodore Roosevelt High School (Bronx, New York), where he focused his attention on improving the students' problem-solving skills and, at the same time, enriching their instruction far beyond what the traditional textbooks offered. During his six-year tenure there, he also developed the school's first mathematics teams (both at the junior and senior levels). He is still involved in working with mathematics teachers and supervisors, nationally and internationally, to help them maximise their effectiveness.

Immediately upon joining the faculty of the City College of New York in 1970 (after having received his master's degree there in 1966), he began to develop in-service courses for secondary school mathematics teachers, including such special areas as recreational mathematics and problem-solving in mathematics. As dean of the City College School of Education for 10 years, his scope of interest in educational issues covered the full gamut of educational issues. During his tenure as dean, he took the School of Education from the bottom of the New York State rankings to the top with a perfect NCATE accreditation assessment in 2009. He also raised more than $12 million from the private sector for educational innovative programmes. Posamentier repeated this successful transition at Mercy College, where he enabled it to become the only college to have received both NCATE and TEAC accreditation simultaneously.

In 1973, Dr. Posamentier received his PhD from Fordham University (New York) in mathematics education and has since extended his reputation in mathematics education to Europe. He has been a visiting professor at several European universities in Austria, England, Germany, the Czech Republic, Turkey and Poland. In 1990, he served as Fulbright Professor at the University of Vienna.

In 1989, he was awarded an Honorary Fellow position at the South Bank University (London, England). In recognition of his outstanding teaching, the City College Alumni Association named him Educator of the Year in 1994 and in 2009. New York City had the day, May 1, 1994, named in his honor by the President of the New York City Council. In 1994, he was also awarded the *Das Grosse Ehrenzeichen für Verdienste um die Republik Österreich* (Grand Medal of Honor from the Republic of Austria), and in 1999, upon approval of Parliament, the President of the Republic of Austria awarded him the title of University Professor of Austria. In 2003, he was awarded the title of *Ehrenbürgerschaft* (Honorary Fellow) of the Vienna University of Technology, and in 2004, he was awarded the *Österreichisches Ehrenkreuz für Wissenschaft & Kunst 1.Klasse* (Austrian Cross of Honor for

Arts and Science, First Class) from the President of the Republic of Austria. In 2005, he was inducted into the Hunter College Alumni Hall of Fame, and in 2006, he was awarded the prestigious Townsend Harris Medal by the City College Alumni Association. He was inducted into the New York State Mathematics Educator's Hall of Fame in 2009, and in 2010, he was awarded the coveted Christian-Peter-Beuth Prize from the Technische Fachhochschule – Berlin. In 2017, Posamentier was awarded *Summa Cum Laude nemmine discrepante* by the Fundacion Sebastian, A.C., Mexico City, Mexico.

He has taken on numerous important leadership positions in mathematics education locally. He was a member of the New York State Education Commissioner's Blue Ribbon Panel on the Math-A Regents Exams and the Commissioner's Mathematics Standards Committee, which redefined the Mathematics Standards for New York State, and he also served on the New York City schools' Chancellor's Math Advisory Panel.

Dr. Posamentier is still a leading commentator on educational issues and continues his long-time passion of seeking ways to make mathematics interesting to teachers, students and the general public – as can be seen from some of his more recent books.

For more information and a list of his publications, see: https://en.wikipedia.org/wiki/Alfred_S._Posamentier.

**Robert Geretschläger** is currently a lecturer at the University of Graz, Austria. He was a teacher of mathematics and geometry, living in Graz, Austria. He retired from his teaching position at Bundesrealgymnasium Keplerstraße in Graz in 2022 after working there for nearly 40 years. He is currently still active as a geometry lecturer in teacher education at the University of Graz, Austria.

Born in Toronto, Canada, in 1957, he moved to Austria in 1972, where he completed his education, certified with a magister degree as a secondary school mathematics teacher with a speciality in geometry, and subsequently earned a doctorate in mathematics from the University of Graz.

For most of his career, he has been actively involved in the organisation of mathematics competitions at all levels of age and complexity. Among other roles in this context, he has been the leader of the Austrian team at the International Mathematical Olympiad since 2007 and the leading organiser of the Mathematical Kangaroo Competition in Austria since its beginning in the mid-1990s. He was also responsible for introducing the *International Mathematical Tournament of the Towns* and the *Mediterranean Mathematics Competition* to Austria.

Internationally, since 2008, he has been a member of the Executive of the *World Federation of Mathematics Competitions* (WFNMC), serving as senior vice-president from 2018 to 2022 and as president since 2022. He is also a long-standing member of the international board of the *Association Kangourou sans Frontières* (AKSF), where he is currently serving as treasurer. He has also lectured at many international venues, having been an invited lecturer in Canada, Australia, the Czech Republic, Switzerland, Colombia, and Germany, among other countries.

He has also been involved in a number of mathematics education projects over the years. Among other things, he was a member of the Didactics Commission of the *Österreichische Mathematische Gesellschaft* (ÖMG) for many years and the leading editor of a series of high school mathematics books. He was actively involved in curriculum development for the Austrian high schools and played an active role in the early phase of introducing the *PISA* study to Austria.

He is the editor and co-author of numerous books on popular mathematics, problem-solving (competition) mathematics and his special research interest, the geometry of origami.

More information is available at his website at www.rgeretschlaeger.com.

# Acknowledgments

It is well known that mathematical problems often lend themselves to alternative solutions. Our goal in this book has been to provide the most elegant and efficient solutions and proofs to some of the most challenging geometric problems. Towards this end, several mathematicians have provided us with some wonderful ideas. These include David Hankin, Moritz Hiebler, Hans Humenberger and Robert Serkey. We thank them for sharing their brilliance with us.

# Contents

# Introduction

Although we are all exposed to geometric principles in early school-ing, not all of us are given the opportunity to truly appreciate the full power and beauty of geometry. Considered either from a purely aesthetic standpoint or as a logical subject, geometry can be consid-ered as a key to the appreciation of mathematics. In the United States, the exposure that students get by studying geometry for one entire year in secondary school is certainly enlightening. However, with so much time spent developing an understanding for the basic concepts and working with the underlying principles of geometry, there is lit-tle time left to truly appreciate some of the wonders that the subject offers. In some countries, the study of elementary geometry has been radically reduced in recent decades, and appreciation of its beauty has suffered accordingly. Even in countries with a strong tradition of teaching geometry in school, the emphasis tends to be more on tech-nical aspects and logic, often to the detriment of the recognition of its visual splendor.

There are many astonishing and surprising results to be found when geometric relationships are explored. Many of these are quite counterintuitive, but astounding, nevertheless. A simple example, sometimes presented in secondary school geometry courses, is the fact that consecutively joining the midpoints of the sides of any ran-domly drawn quadrilateral always yields a parallelogram. There are countless such unexpected curiosities that result from rather simple geometric formations. This book will take motivated readers on a journey through many previously unexplored relationships that will leave them in total amazement. Also, to enlighten the readership, Euclidean proofs are provided for all of these astonishing results. These proofs are presented in as reader-friendly a fashion as possible. Of course, many readers may be quite content just to marvel at the results without concerning themselves with the proofs. This is fine; as understanding the proofs is in no way required to appreciate these lovely properties.

This is the third and final book of a three-volume series exhibiting the splendors of geometry, and it focuses on circle relationships. The first book in the series concentrates on triangles and the second book on quadrilaterals. In order to benefit the most from reading this book, the reader should have a reasonable familiarity with the basic concepts taught in secondary school geometry. When there are geometric theorems used in some of the proofs, which may be considered beyond the scope of secondary school geometry, we present them in a Toolbox section which will elucidate them for the general readership. One should not think that these Toolbox items are all beyond the high school curriculum, as a good portion of this section reviews many of the basics from high school geometry courses as well. In short, the Toolbox provides all the necessary equipment a reader will need to understand the proofs offered in the second section of this book.

To make your journey through the book as enjoyable and as entertaining as possible, the organization of the individual examples is rather randomly dispersed, although there are several related sequences intentionally organized to show the development of certain concepts. Some of these sequential units might deal with collinearity of points, concurrency of lines, concyclic points, and other topics typically not emphasized in the secondary school geometry curriculum.

The examples selected throughout this book are quite uncommon and have either been self-created or else presented in geometry books over the past two centuries. Readers are advised to first read and truly appreciate the amazing relationships that each of these examples presents. As a next step, an interested reader might focus on proving these astonishing curiosities, especially after experiencing a genuine appreciation of its wonders. To enhance the appreciation of these, we have done our best to present all proofs and enhancements in a reader-friendly fashion.

We invite you now to embark on this marvelous journey through many hidden features of quadrilateral-related geometry that expose the power and beauty of geometry as well as the logic that supports it. Enjoy!

# Circle Curiosities: Introducing the Circle

To begin our exploration of the countless geometric surprises that the circle offers, we should quickly review some of the circle's most basic features. The first few Curiosities presented here will be familiar to most readers, but they are nevertheless fascinating and it's important to understand why they are universally true.

Clearly, the most basic aspect of a circle is its definition. A circle $c$ is composed of all points $P$ (in a plane) whose distance from a given point $O$ (called the *center* of the circle) is equal to a specific given value $r$ (called the *radius* of the circle), as shown in Figure A.

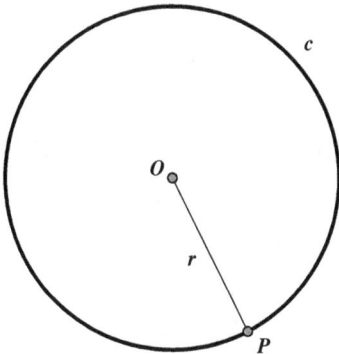

Figure A

This definition immediately leads us to the first Curiosity, and it will be worth your while to consider the reason why it's an immediate consequence of the definition.

### Curiosity 1. A Triangle Defines a Circle

Any three non-collinear points determine a unique circle. In Figure 1, we see three points *A*, *B*, and *C*. No matter where we choose these points, as long as they do not lie on a common line, there exists a unique circle *c* containing all three of these points.

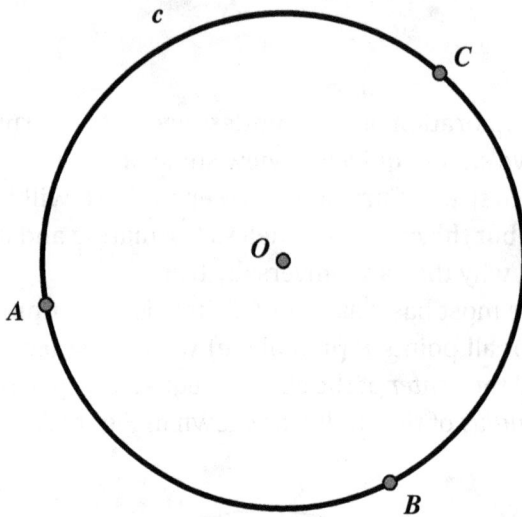

Figure 1

### Curiosity 2. A Circle Tangent is Perpendicular to the Radius in the Point of Tangency

A *tangent* of a circle is a line that has exactly one point in common with the circle. As shown in Figure 2, the radius drawn to the point *A* of contact of a tangent line to a circle *c* is perpendicular to the tangent line in *A*.

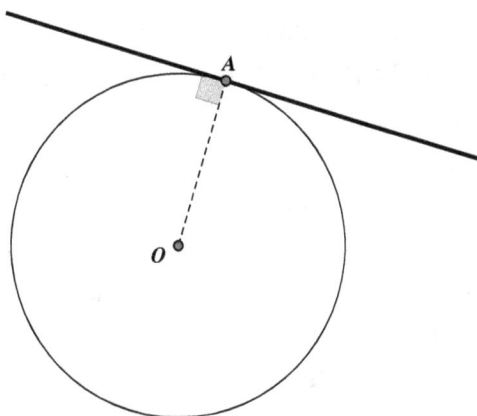

Figure 2

## Curiosity 3. Circle Tangents from a Common Point are of Equal Length

Two tangents drawn from an external point to the same circle are equal in length. In Figure 3, we see tangents *PA* and *PB* of circle *c*, and we find that *PA = PB*.

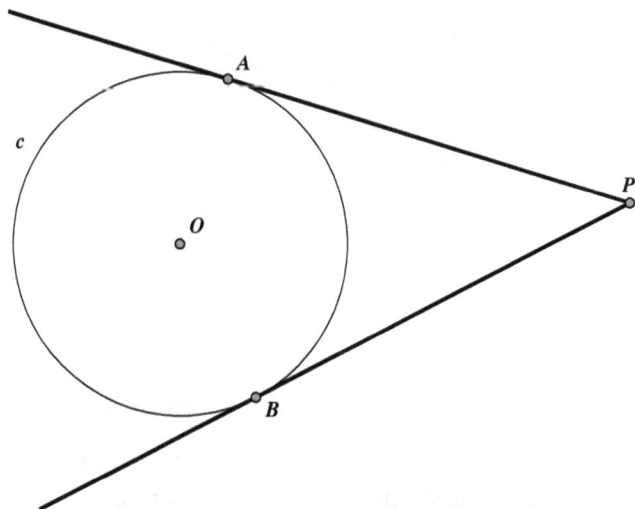

Figure 3

## Curiosity 4. The Angle Between Circle Tangents from a Common Point is Supplementary to its Intercepted Arc

The measure of an angle formed by two tangents from a common external point to a circle is equal to one-half the difference of the measures of the intercepted arcs. That is, in Figure 4, $\angle P = \frac{1}{2}\left(\text{large}\,\widehat{AB} - \text{small}\,\widehat{AB}\right)$. Also, the measure of the angle formed by two tangents from an external point is supplementary to the measure of the closer intercepted arc, or $\angle P = 180° - \text{small}\,\widehat{AB}$. This can also be written as $\angle APB = \frac{1}{2}\left((360° - \angle BOA) - \angle BOA\right) = 180° - \angle BOA$.

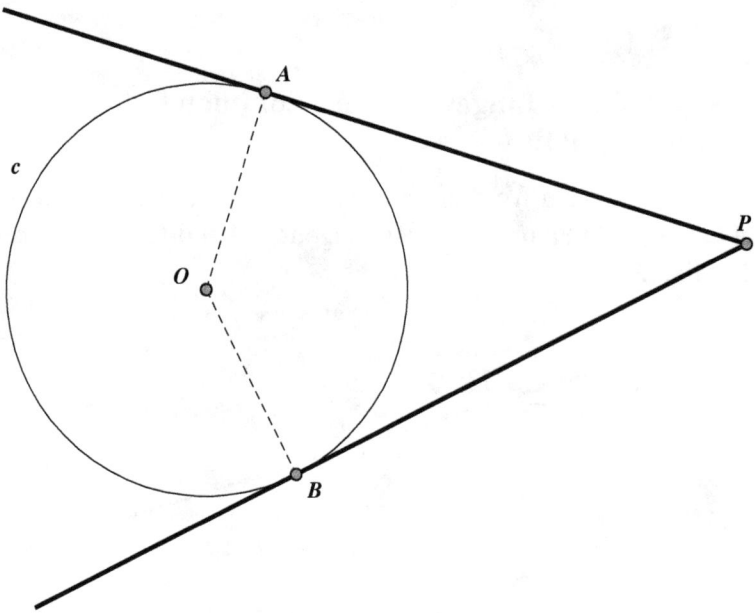

Figure 4

## Curiosity 5. Angle Between a Circle Tangent and a Chord

The measure of an angle formed by a tangent and a chord of a circle is one-half the measure of its intercepted arc. That is, in Figure 5, with point $P$ on the tangent of circle $c$ in $B$, we have $\angle PBA = \frac{1}{2} \cdot \angle BOA = \frac{1}{2}\,\widehat{AB}$.

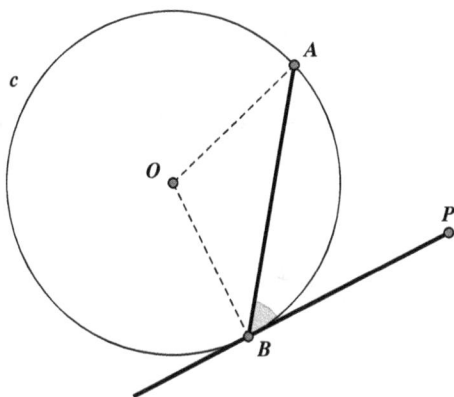

Figure 5

## Curiosity 6. The Angle Formed by Two Chords of a Circle

When two chords intersect in the interior of a circle, the product of the lengths of the segments of one chord equals the product of the lengths of the other chord. For the two chords $AB$ and $CD$ intersecting at the point $P$ in Figure 6, the following holds: $AP \cdot BP = CP \cdot DP$. Furthermore, the measure of the angle formed by two chords intersecting at a point in the interior of the circle is one-half the sum of the measures of the arcs intercepted by the angle and its vertical angle. We can write this as $\angle CPA = \frac{1}{2}(\overset{\frown}{AC} + \overset{\frown}{BD})$ or $\angle CPA = \frac{1}{2}(\angle COA + \angle DOB)$.

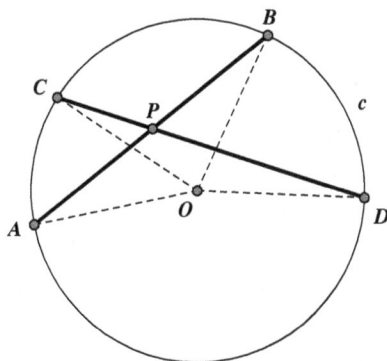

Figure 6

## Curiosity 7. Surprising Supplementary Angles

In Figure 7, a circle with center $O$ has chords $AB$ and $CD$ perpendicular at point $P$. Surprisingly, we find that angles $\angle AOD$ and $\angle BOC$ are supplementary, that is, $\angle AOD + \angle BOC = 180°$.

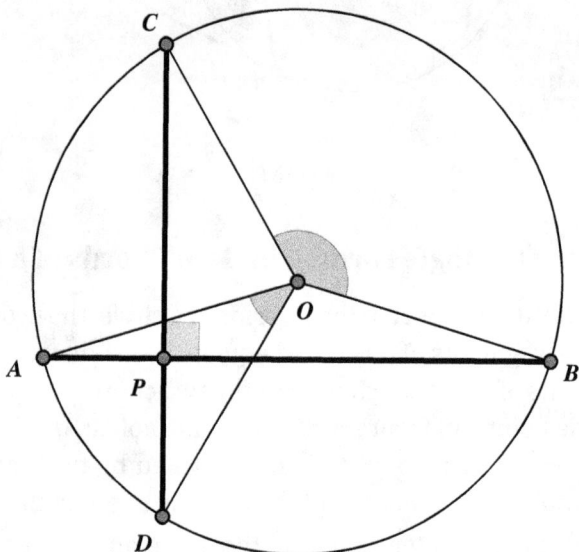

Figure 7

## Curiosity 8. The Angle Formed by a Tangent and a Secant

If a secant and a tangent segment are drawn to the same circle from an external point, then the tangent segment is the mean proportional between the secant and its external segment. In Figure 8, tangent $PA$ is the mean proportional between $PB$ and $PC$, so that $\frac{PB}{PA} = \frac{PA}{PC}$. The measure of an angle formed by a secant and a tangent to a circle from an external point is equal to one-half the difference of the measures of the intercepted arcs. We can write this as $\angle APC = \frac{1}{2}(\widehat{AC} - \widehat{AB})$ or $\angle APC = \frac{1}{2} \cdot (\angle AOC - \angle BOA)$.

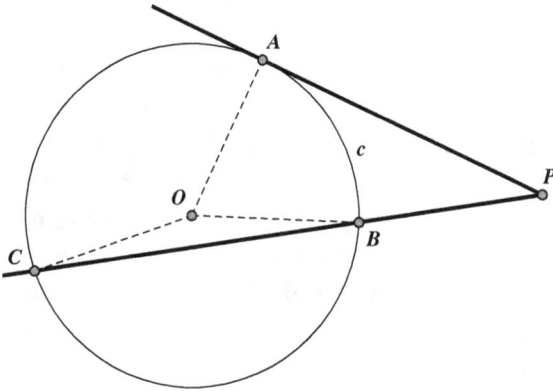

Figure 8

## Curiosity 9. The Angle Formed by Two Secants

When two secants are drawn to the same circle from an external point, the product of the lengths of one secant and its external segment equals the product of the lengths of the other secant and its external segment. In Figure 9, for the two secants *PBA* and *PDC*, the following is true: $AP \cdot BP = CP \cdot DP$. The measure of an angle formed by two secants of the circle is equal to one-half the difference of the measures of the intercepted arcs. We can write this as $\angle CPA = \frac{1}{2}(\widehat{AC} - \widehat{DB})$ or $\angle CPA - \frac{1}{2} \cdot (\angle COA - \angle BOD)$.

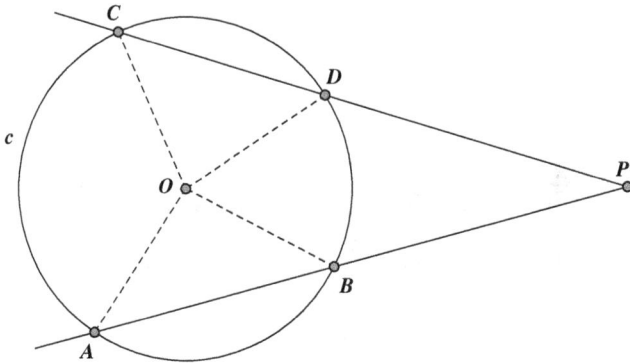

Figure 9

## Curiosity 10. Trisecting an Angle Using a Circle

One of the three famous unsolvable problems of antiquity is to trisect a general angle using only a straightedge and compasses. We can, however, achieve an angle trisection in the following way. In Figure 10, the goal is to trisect $\angle ADE$, and it will turn out that $\angle ABE$ is exactly one-third as large. To find point $B$, we first construct a circle with center $D$ and radius $CD$. We then construct line segment $CB$ to meet the extension of $AD$ at point $B$, so that $CD = BC$. Furthermore, the extension of $BC$ should intersect the circle at point $E$. This then results in the desired property $\angle ABE = \frac{1}{3} \cdot \angle ADE$.

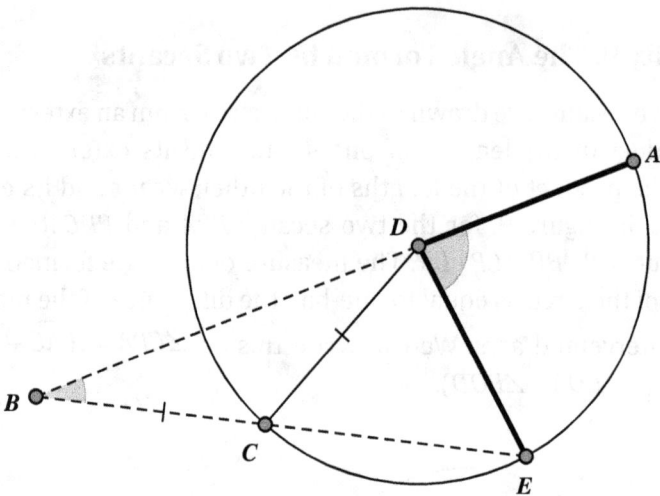

Figure 10

## Curiosity 11. A Rather Surprising Equality

Here is a situation that is simple and simply amazing! The points $A$ and $B$ are on circle $O$. Tangents of equal length are constructed at the points $A$ and $B$, as we see in Figure 11, where $AL = BK$. When $AB$ is extended to intersect $LK$ at point $P$, unexpectedly, we find that $KP = LP$.

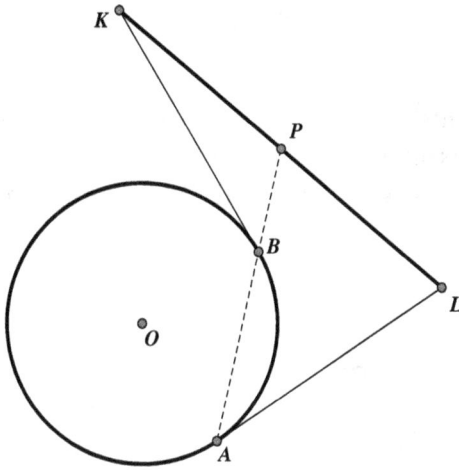

Figure 11

## Curiosity 12. A Strange Appearance of Parallel Lines

In Figure 12, line *t* is the tangent of circle *c* at point *A*. A randomly selected line *l* intersects circle *c* at points *B* and *C*, and tangent *t* intersects circle *c* at point *T*. Point *D* is the midpoint of arc *BC*, not containing point *A*. When the perpendicular to *AD*, which intersects *AD* at point *X*, is extended to point *R* so that $TX = RX$, unexpectedly, we find that line *AR* is parallel to the randomly chosen line *l*.

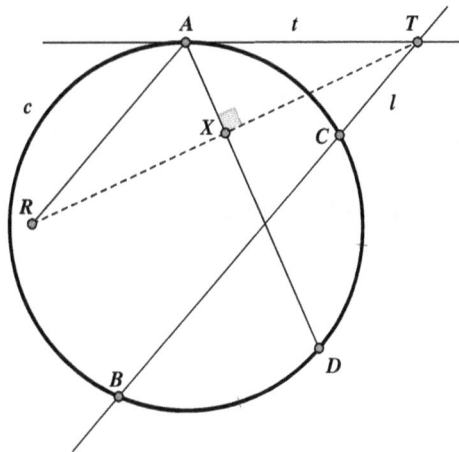

Figure 12

## Curiosity 13. Unexpected Complementary Angles in a Circle

In Figure 13, points $D$ and $C$ are on circle $O$, on the same side of diameter $AOB$. Point $E$ is on the opposite side of diameter $AOB$. Unexpectedly, we find that the angles in $D$ and $C$ are complementary, that is, $\angle ADE + \angle ECB = 90°$.

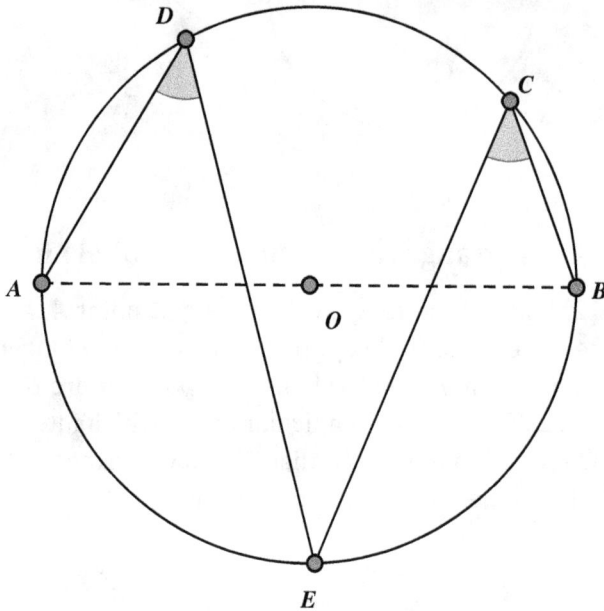

Figure 13

## Curiosity 14. Surprising Squares Equality in a Circle

In Figure 14, we have four chords $AB$, $BC$, $CD$, and $DE$ drawn in a circle so that $\angle ABC = \angle DCB = \angle CDE = 45°$. Quite unexpectedly, we find that $AB^2 + CD^2 = BC^2 + DE^2$.

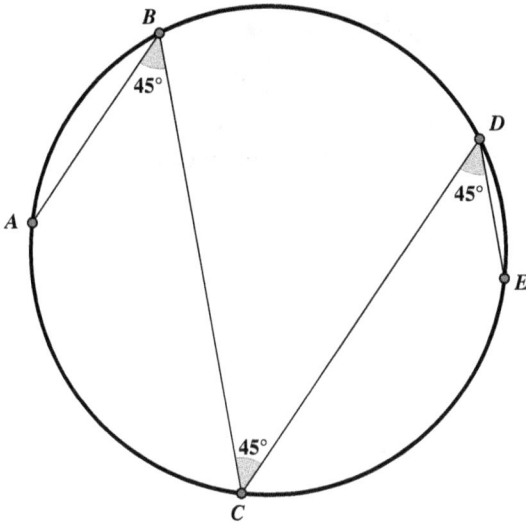

Figure 14

## Curiosity 15. A Surprise of a Circle Presenting a Familiar Triangle

By drawing a chord one-half the length of the diameter of the circle and parallel to it, we can create a triangle in which the longest side is twice as long as the shortest side. In Figure 15, we are given a circle with center $O$ and diameter $AB$. When we draw chord $DE$ parallel to diameter $AB$ with $DE = \frac{1}{2} \cdot AB$, we obtain the point $C$, where $AD$ intersects the tangent at point $B$. Unexpectedly, we then have $AC = 2AB$.

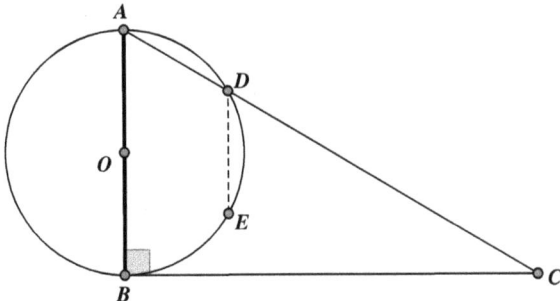

Figure 15

## Curiosity 16. A Surprising Perpendicularity

In Figure 16, line $AC$ is a chord of circle $O$, whose diameter is $AOB$. Line $AT$ bisects angle $\angle CAB$ and intersects circle $O$ at point $T$. Surprisingly, we find that the tangent to circle $O$ at point $T$ is perpendicular to the extension of $AC$.

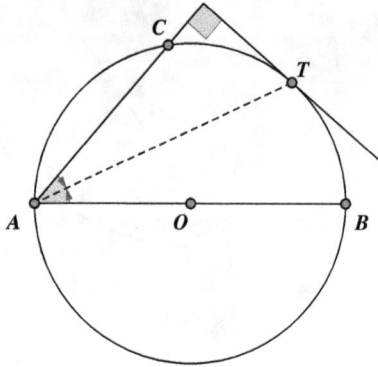

Figure 16

## Curiosity 17. Unexpected Concyclic Points

When a randomly drawn chord $AB$ is drawn in circle $c$ with center $O$, as shown in Figure 17, the midpoint $P$ of arc $AB$ produces four concyclic points in a very neat way. When two lines are drawn at random through point $P$, they intersect $AB$ at points $G$ and $H$, and intersect circle $c$ at points $C$ and $D$, respectively. Regardless of where these chords are drawn, the points $C$, $D$, $H$, and $G$ will always be concyclic.

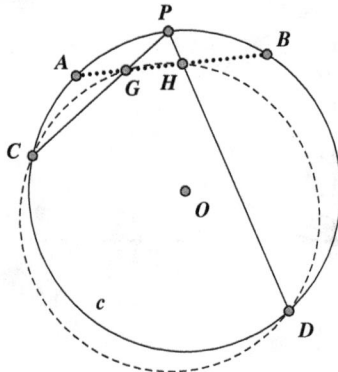

Figure 17

## Curiosity 18. Circle Chords Generate a Constant Angle Sum

In Figure 18, we are given a circle with center $O$ and two points $A$ and $B$ on the circumference. Point $C$ is chosen at random on an arc $AB$. When points $D$ and $E$ are randomly chosen with point $D$ on the arc $AC$ (which does not contain point $B$) and point $E$ on the arc $CB$ (which does not contain point $A$), we find that the sum $\angle CDA + \angle BEC$ is the same regardless of the placement of points $C$, $D$, and $E$ on their respective arcs.

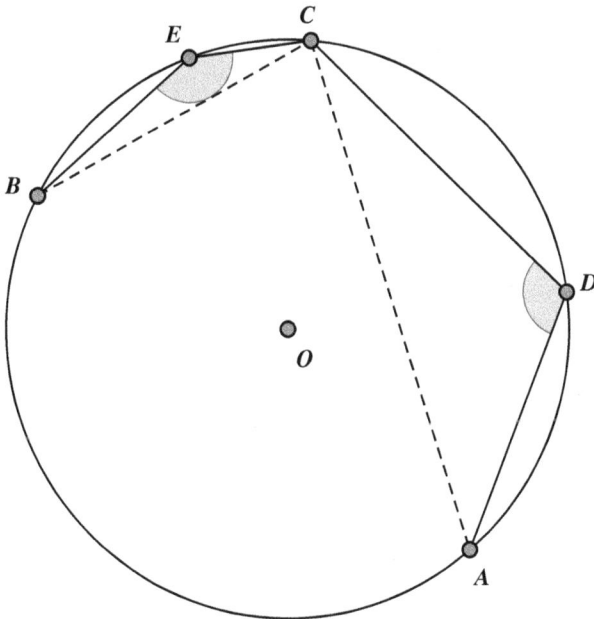

Figure 18

## Curiosity 19. A Circle Helps Divide a Line Segment into Two Equal Parts

In Figure 19, point $E$ is any point along diameter $AB$ of circle $O$. A perpendicular to the diameter is erected at point $E$, intersecting circle $O$ at point $D$. We erect a perpendicular to diameter $AB$ at point $B$ and extend $AD$ to meet that perpendicular at point $F$. From the circle's center $O$, a parallel to $AF$ intersects $FB$ at point $C$. The line $AC$ then divides $DE$ into two equal parts so that $DP = EP$.

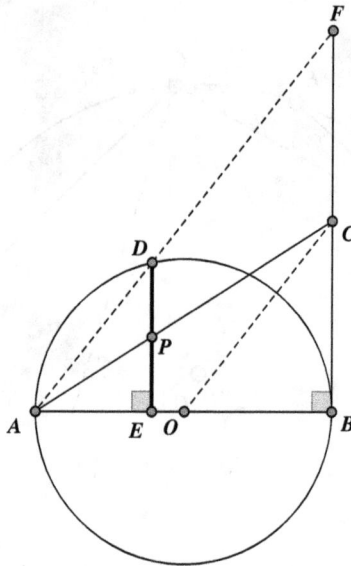

Figure 19

## Curiosity 20. The Famous Butterfly Relationship

Four well-placed chords can partition a fifth chord of a circle into equal parts. In Figure 20, point $M$ is the midpoint of a chord $AB$ of a given circle. Chords $CD$ and $EF$ are randomly selected and intersect at point $M$. Point $P$ is the intersection of $AB$ and $CF$, and point $Q$ is the intersection of $AB$ and $DE$. Quite unexpectedly, we find that $MP = MQ$. This curiosity is often referred to as the *Butterfly Problem* because of the shape of the figure produced by the four chords.

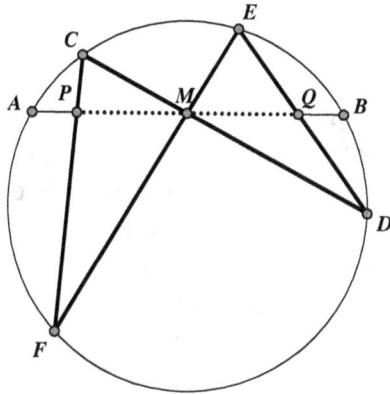

Figure 20

## Curiosity 21. A Circle Unexpectedly Produces Parallel Lines

In Figure 21, circle $O$ has diameter $AOB$ and point $C$ on its circumference. Point $P$ is the point at which the tangents of the circle in $B$ and $C$ intersect. Unexpectedly, it turns out that $AC$ is parallel to $OP$.

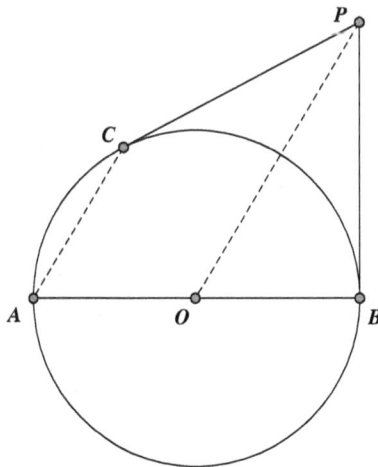

Figure 21

## Curiosity 22. The Unexpected Constant for Any Point on a Circle

Two points on the diameter of the circle, which are equidistant from the center, have the interesting property that the sum of the squares of their distances to any point on the circumference of the circle is the same. For example, in Figure 22, points $C$ and $D$ on diameter $AOB$ are equidistant from the center $O$, such that $OC = OD$. When point $P$ is selected anywhere on the circumference of the circle, the value of $PC^2 + PD^2$ is independent of the choice of $P$.

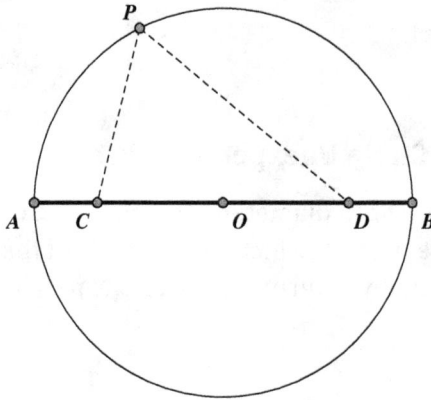

Figure 22

## Curiosity 23. Circle Tangents Generate Perpendicular Lines and Proportional Segments

In Figure 23, we are given a circle with center $O$ and diameter $AB$. The circle tangents $AD$ and $BC$ are parallel and intersected in $D$ and $C$, respectively, by a tangent line to the circle at point $E$. We then find that $\angle COD = 90°$. Furthermore, the radius of the circle is the mean proportional between the two parallel tangent segments, which gives us the proportion $\frac{OE}{BC} = \frac{AD}{OE}$.

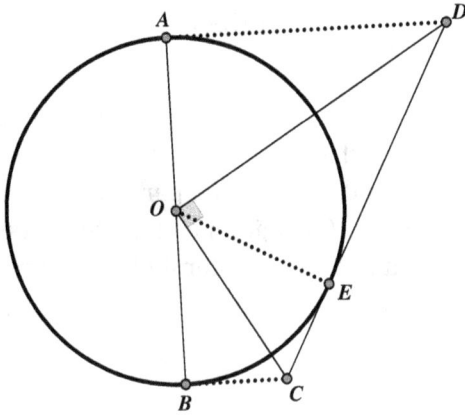

Figure 23

## Curiosity 24. An Unexpected Circle Diameter

In Figure 24, we are given points $A$, $B$, $C$, and $D$ on a circle with center $O$. The angle bisector of $\angle ABC$ intersects the circle a second time at point $X$, and the angle bisector of $\angle CDA$ intersects the circle a second time at point $Y$. Unexpectedly, $XY$ turns out to be a diameter of the original circle $O$.

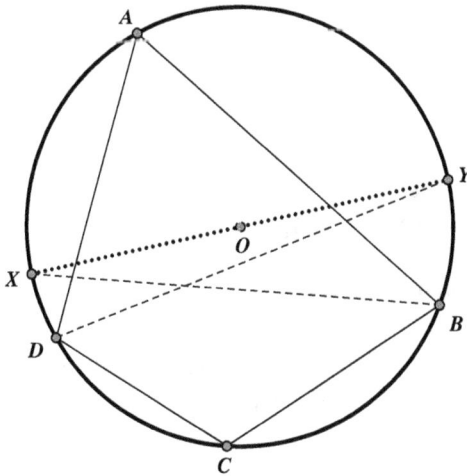

Figure 24

### Curiosity 25. An Unexpected Rectangle in a Circle

We are given a circle with center $O$ and four points $A$, $B$, $C$, and $D$, arranged as shown in Figure 25, where $AC$ is the diameter of the circle. Point $E$ also lies on the circle, such that $BE \perp AC$, and point $F$ is the intersection of $BE$ and $AC$. The line parallel to $BD$ through point $F$ intersects the extension of $CD$ at point $P$ and intersects line $AD$ at point $Q$. We surprisingly find that quadrilateral $EQDB$ is a rectangle.

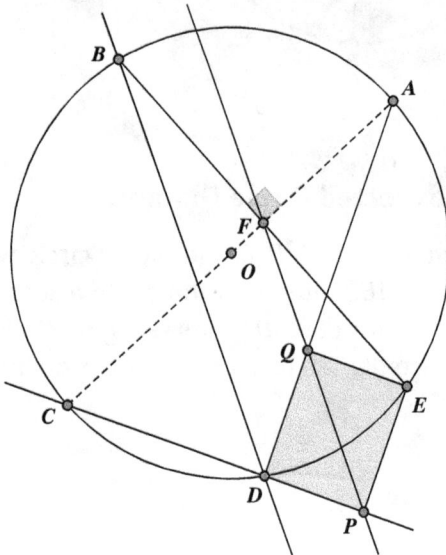

Figure 25

### Curiosity 26. Parallel Angle Bisectors

Four points $A$, $B$, $C$, and $D$ lie on a circle, as shown in Figure 26. Point $E$ is the intersection of diagonals $AC$ and $BD$, and point $F$ is the intersection of the extensions of $AB$ and $DC$. Surprisingly, we find that the angle bisectors of $\angle DEA$ and $\angle DFA$ are parallel.

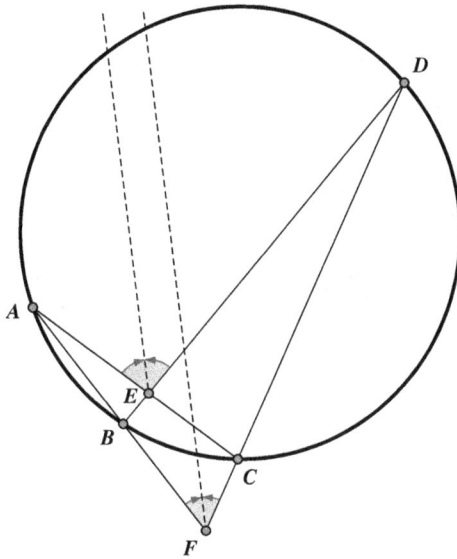

Figure 26

## Curiosity 27. Noteworthy Perpendiculars Generated by a Semicircle

We are given two points $P$ and $Q$ on a semicircle with diameter $AB$, as shown in Figure 27. Point $M$ is the midpoint of $PQ$. When we determine the point $C$ on the extension of $AM$ such that $MA = MC$, we find that lines $PQ$ and $BC$ are perpendicular.

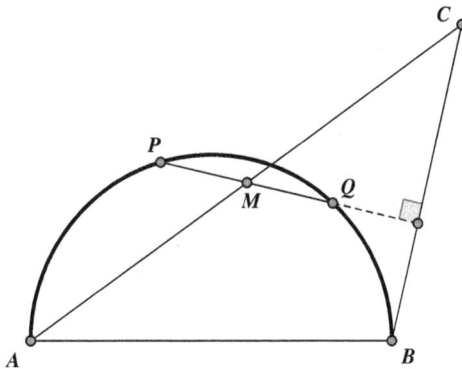

Figure 27

## Curiosity 28. An Unexpected Angle Bisector in a Semicircle

We are given two points $P$ and $Q$ on a semicircle with diameter $AB$, as shown in Figure 28. Lines $AQ$ and $BP$ intersect at point $R$, and point $S$ is on $AB$ so that $RS \perp AB$. We then find that $RS$ is the angle bisector of $\angle QSP$.

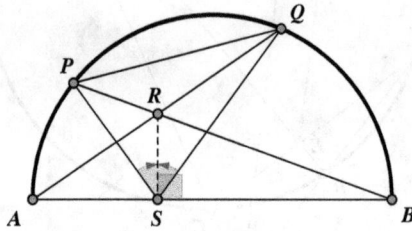

Figure 28

## Curiosity 29. Tangent Circles on a Semicircle

We are given a point $P$ on a semicircle with diameter $AB$, as shown in Figure 29, as well as a point $Q$ in the interior of the semicircle. The line $AQ$ extended intersects the semicircle at point $C$ and intersects $BP$ at point $E$, and the line $BQ$ extended intersects the semicircle at point $D$ and intersects $AP$ at point $F$.

Quite unexpectedly, we find that the circumcircles of triangles $PDF$ and $PEC$ are tangent at point $P$.

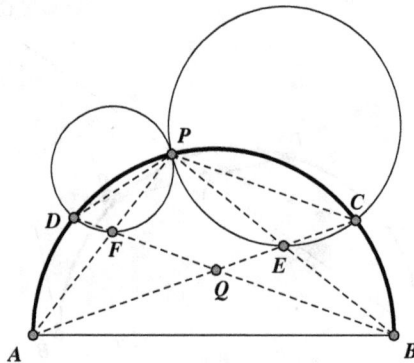

Figure 29

## Curiosity 30. A Point on the Circumcircle of an Isosceles Triangle

Points on a circle also provide us with some very interesting relation-ships. Take, for example, a point $P$ on the circumscribed circle of isos-celes triangle $ABC$, as shown in Figure 30. We have $AB = AC$, and point $P$ lies on the arc $BC$ of the circumcircle not containing $A$. The following ratio then holds: $\frac{AC}{BC} = \frac{PA}{PB+PC}$.

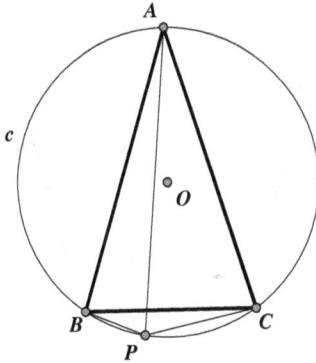

Figure 30

## Curiosity 31. A Point on the Circumcircle of a Square

Curiosity 30 can be extended to the circumcircle of a square. In Figure 31, we are given a circle circumscribed about the square $ABCD$ and a point $P$ on the arc $BC$ of this circle, not containing $A$ and $D$. In this configuration, we have $\frac{PA+PC}{PB+PD} = \frac{PD}{PA}$.

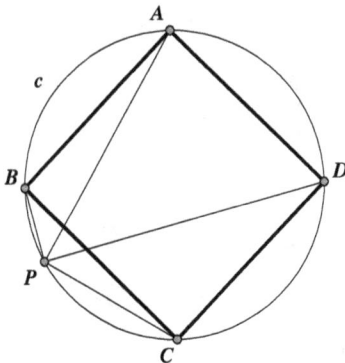

Figure 31

## Curiosity 32. A Point on the Circumcircle of a Regular Pentagon

When a point $P$ is on the circumcircle $c$ of a regular pentagon $ABCDE$, as shown in Figure 32, another unexpected relationship evolves. Here, point $P$ is on the small arc $BC$ of the circle $c$, and we then have $PA + PD = PB + PC + PE$.

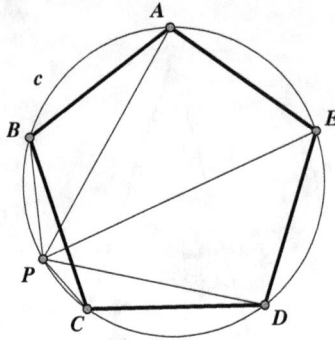

Figure 32

## Curiosity 33. A Point on the Circumcircle of a Regular Hexagon

We can expand upon Curiosity 32 by replacing the pentagon with a hexagon. When we choose a point $P$ on the small arc $BC$ of the circumscribed circle of a regular hexagon $ABCDEF$, as shown in Figure 33, we find that $PE + PF = PA + PB + PC + PD$.

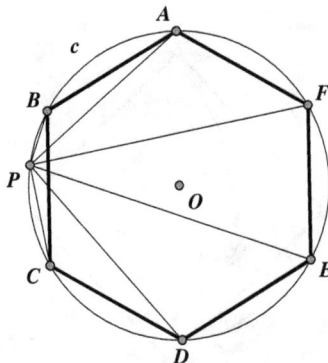

Figure 33

## Curiosity 34. The Tangent to the Circumscribed Circle of an Equilateral Triangle

A tangent line randomly drawn to the circumscribed circle of an equilateral triangle provides a rather unusual relationship. In Figure 34, point $P$ is randomly selected on the circumscribed circle of equilateral triangle $ABC$. From each of the vertices of the triangle, a perpendicular line is a drawn to the line tangent to the circle at point $P$. We find the sum of these three perpendiculars, namely, $AD$, $BE$, and $CF$, is equal to twice the length of the triangle's altitude $AM$. Symbolically, that is $AD + BE + CF = 2AM$.

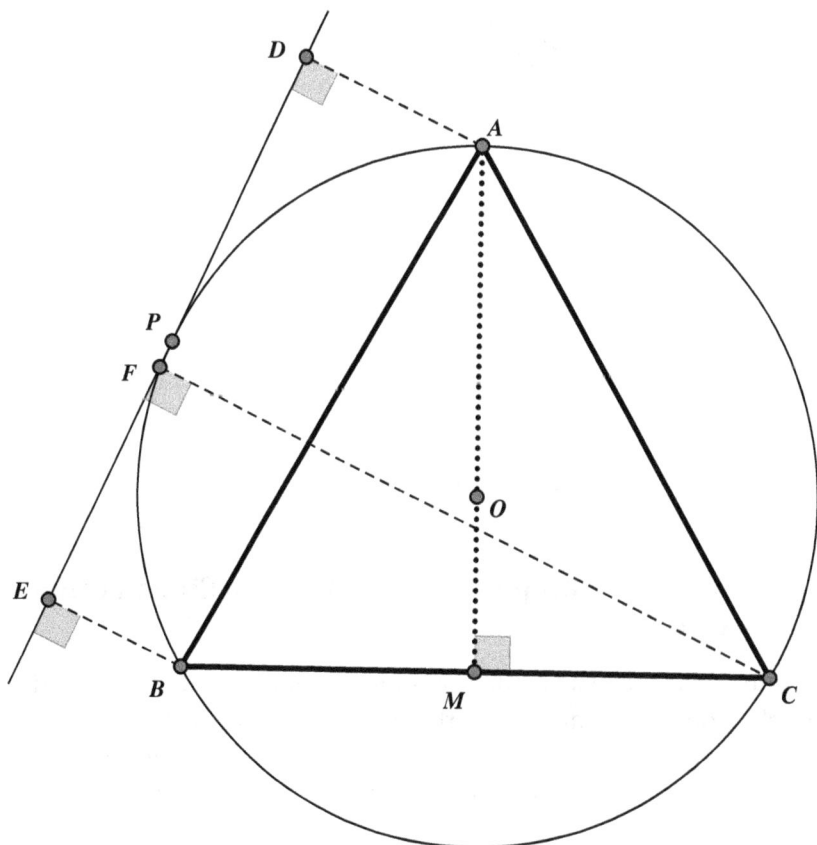

Figure 34

## Curiosity 35. Surprising Concyclic Points

In Figure 35, we are given a triangle *ABC* with points *D* and *E* on sides *AB* and *AC*, respectively, such that *DE*∥*AB*. A circle *c* is drawn containing points *D* and *E*, tangent to *AC* at point *E*, and this circle intersects side *BC* a second time at point *F*. Unexpectedly, we then find that points *A*, *B*, *F*, and *E* are concyclic.

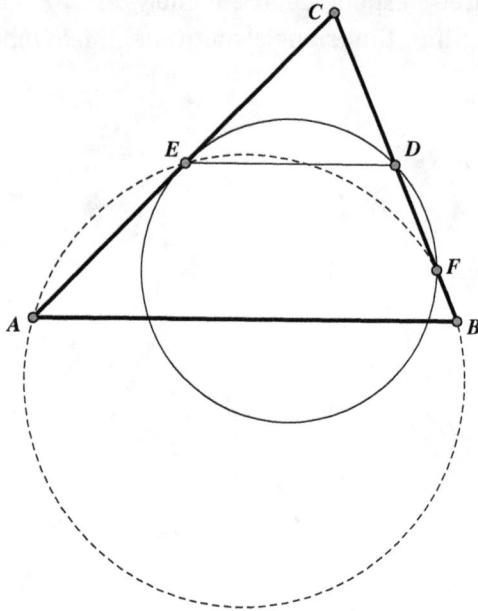

Figure 35

## Curiosity 36. The Midpoints of the Arcs of a Circle Produce Unexpected Parallels

In Figure 36, we consider three points *A*, *B*, and *C* randomly placed on a circle. Points *P*, *Q*, and *R* are the midpoints of arcs *BC*, *CA*, and *AB*, respectively. Lines *AB* and *PR* intersect at point *D* and lines *AC* and *PQ* intersect at point *E*. We then find that line *DE* is parallel to the line *BC*.

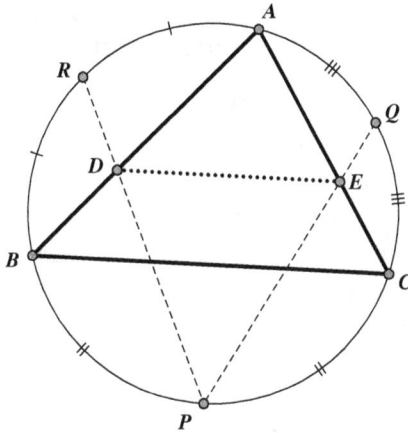

Figure 36

## Curiosity 37. The Circumscribed Circle and the Altitudes of a Triangle Reveal an Unusual Equality

The feet of the altitudes of a randomly selected acute triangle and the triangle's circumcircle surprisingly determine two lines of equal length. In Figure 37, points $E$ and $F$ are the feet of the altitudes from vertices $B$ and $C$ of triangle $ABC$. The extension of line $EF$ intersects the circumcircle at points $P$ and $Q$. Quite unexpectedly, we find that $AP = AQ$.

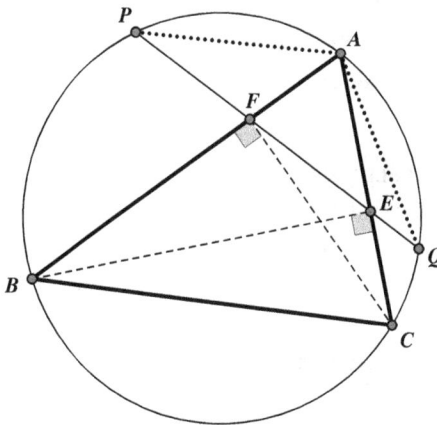

Figure 37

## Curiosity 38. The Unexpected Relationship Between the Midpoint of the Side of an Acute Triangle, its Circumcenter, and its Orthocenter

The distance from the center $O$ of the circumcircle of a triangle to a side of the triangle has an unexpected relationship with the distance between the triangle's orthocenter and the opposite vertex. We see this illustrated in Figure 38, where we are given a triangle $ABC$ with circumcenter $O$ and orthocenter $H$. Point $M$ is the intersection of the perpendicular from point $O$ to side $BC$ of triangle $ABC$. Curiously, we find that $AH = 2OM$.

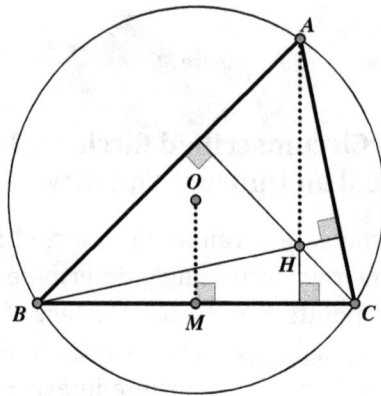

Figure 38

## Curiosity 39. An Unexpected Concurrency with the Circumscribed Circle of the Triangle

The line containing the orthocenter and the midpoint of a side of a given triangle intersects the circumscribed circle of the triangle at the endpoint of the diameter from the opposite vertex. This can be seen in Figure 39, where we are given triangle $ABC$, its circumcircle centered at $O$, and its orthocenter $H$. Point $N$ is the midpoint of side $AB$, and $HN$ extended intersects the circumcircle at point $P$. We then find that $CP$ is a diameter of the circumcircle.

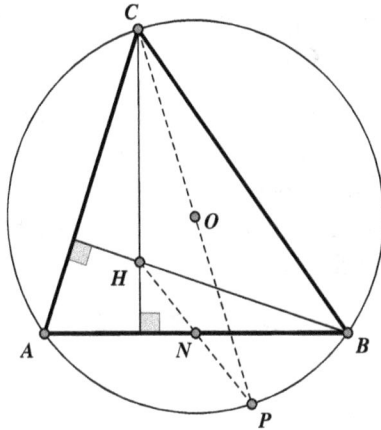

Figure 39

## Curiosity 40. An Unexpected Angle Bisector Hidden Within the Circumscribed Circle

In Figure 40, the circumscribed circle about triangle *ABC* has tangents *AN*, *NBM*, and *CM*. Also, *BP* is perpendicular to *AC* at point *P*. Unexpectedly, *BP* bisects angle *NPM*.

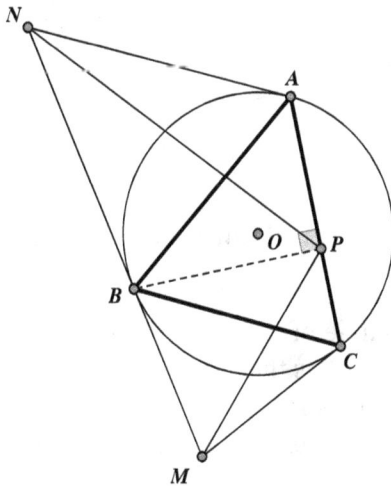

Figure 40

## Curiosity 41. Surprising Concyclic Points Generated by a Triangle and its Circumscribed Circle

In Figure 41, point $P$ is a randomly selected point in the interior of triangle $ABC$. Lines $AP$, $BP$, and $CP$ intersect the circumscribed circle of $ABC$ at points $Q$, $R$, and $S$, respectively. Point $D$ is on $AQ$ and point $F$ is on $BR$ so that $DE\|AB$. Also, point $F$ is on $CS$, so that $DF\|AC$. Quite unexpectedly, we find that the points $S$, $F$, $E$, and $R$ lie on the same circle.

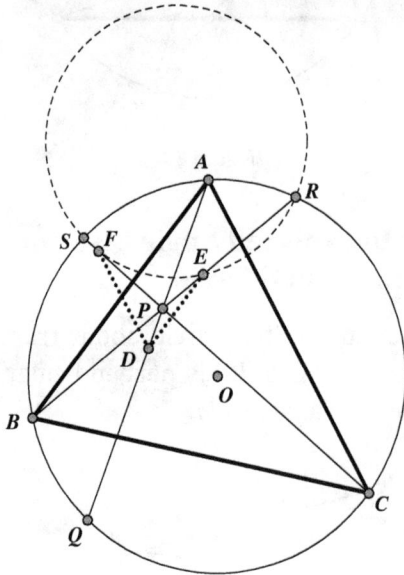

Figure 41

## Curiosity 42. Unexpected Concyclic Points

Consider triangle $ABC$, shown in Figure 42, where $O$ is the center of the circumscribed circle of triangle $ABC$, and $H$ is the orthocenter of the triangle. Segment $AD$ is the altitude to side $BC$. The perpendicular bisector of $AO$ intersects $BC$ extended at point $E$. Quite surprisingly, the circle containing points $A$, $D$, and $E$ also contains the midpoint $M$ of segment $HO$.

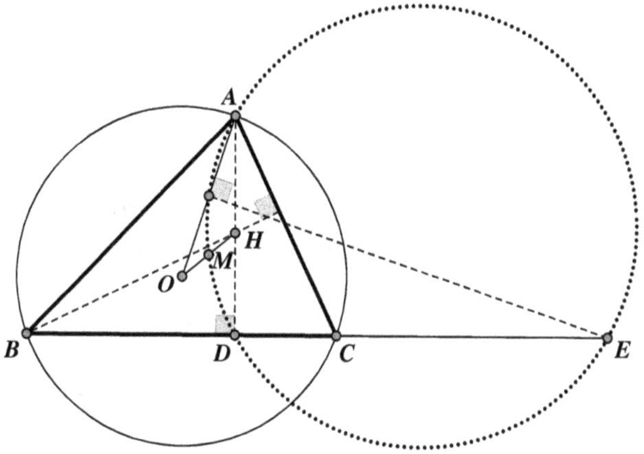

Figure 42

## Curiosity 43. The Orthocenter and Two Vertices of a Triangle Generate a Circle Equal to the Circumcircle

In Figure 43, point $H$ is the orthocenter of triangle $ABC$ and point $O$ is its circumcenter. Surprisingly, we find that any circle containing two vertices of triangle $ABC$ and the point $H$ is equal in size to the circumcircle of triangle $ABC$. In the figure, we see this illustrated by the circumcircle of $BCH$, with center $Q$, which is equal in area to the circumcircle of triangle $ABC$.

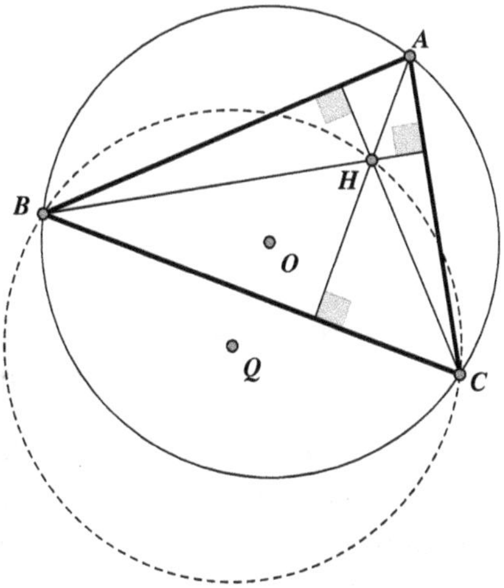

Figure 43

## Curiosity 44. The Appearance of an Unexpected Isosceles Triangle

We are given a triangle $ABC$, as shown in Figure 44. The tangents from point $P$ to the circumcircle $O$ of triangle $ABC$ intersect the circle at points $A$ and $C$. The line through point $P$ parallel to side $AB$ intersects side $BC$ at point $Q$. We find that triangle $ABQ$ is isosceles, with $AQ = BQ$.

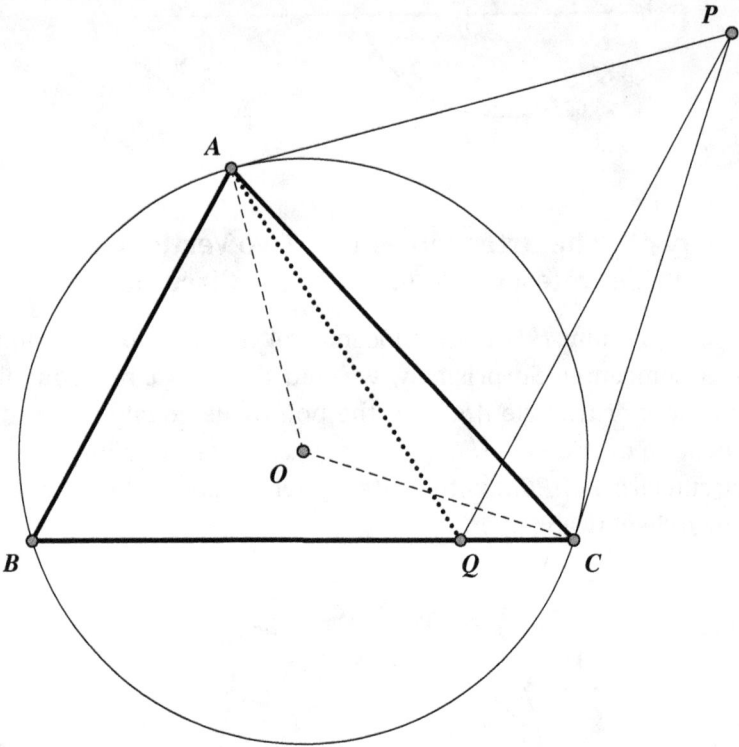

Figure 44

## Curiosity 45. Perplexing Parallels from Concyclic Points

Points $A$, $B$, $C$, $D$, and $E$ lie on a circle with center $O$, with $AB = AE$, as shown in Figure 45. Point $X$ is the intersection of lines $AD$ and $CE$, and point $Y$ is the intersection of lines $AC$ and $BD$. We unexpectedly find that lines $BE$ and $XY$ are parallel.

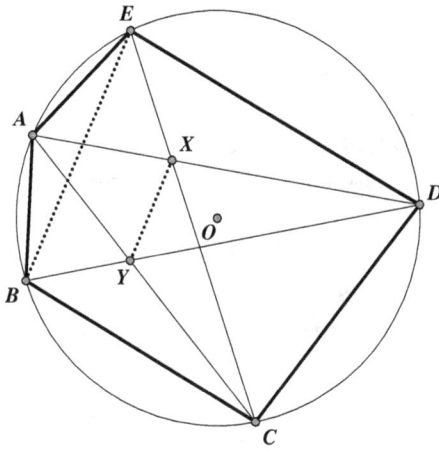

Figure 45

## Curiosity 46. Tangents to the Circumcircle of a Triangle

In Figure 46, the feet of the altitudes *AD*, *BE*, and *CF* of triangle *ABC* determine the orthic triangle *DEF*. There is an interesting connection between the sides of the orthic triangle and the circumcircle of *ABC*. The tangents to the circumcircle at the vertices of the triangle are parallel to the sides of the orthic triangle. In the figure, we see that the tangent at vertex *A* is parallel to *EF*, the tangent at vertex *B* is parallel to *FD*, and the tangent at vertex *C* is parallel to *DE*.

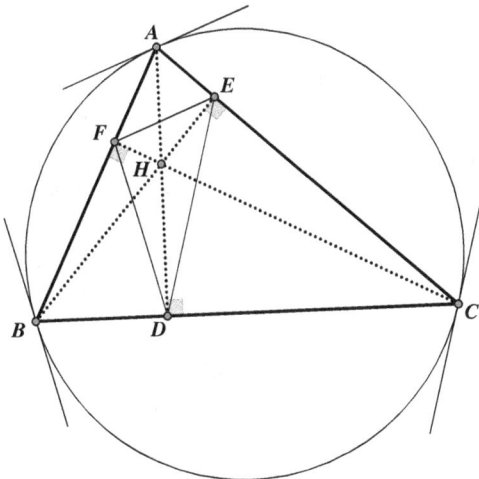

Figure 46

## Curiosity 47. Tangents to the Nine-Point Circle

In this Curiosity, we consider an interesting property of the famous Nine-Point Circle of a triangle (see Toolbox) that is closely related to Curiosity 46. In Figure 47, we see a triangle *ABC*, its orthocenter *H*, and its Nine-Point Circle with center *R*. We recall that the eponymous nine points of the Nine-Point Circle are the midpoints *K*, *M*, and *N* of the triangle sides, the feet *D*, *E*, and *F* of the triangle altitudes, and the midpoints *X*, *Y*, and *Z* of segments *AH*, *BH*, and *CH*, respectively.

Furthermore, as in Curiosity 46, we are given the altitudes *AD*, *BE*, and *CF* of triangle *ABC*, which determine the orthic triangle *DEF*. Astoundingly, the tangents to the Nine-Point Circle at the midpoints of the sides of the triangle are parallel to the sides of the orthic triangle. In Figure 47, we see that the tangent at point *K* is parallel to *EF*, the tangent at point *M* is parallel to *FD*, and the tangent at point *N* is parallel to *DE*. Combining Curiosities 46 and 47, we see that the tangents of the circumcircle of *ABC* at its vertices are also parallel to the tangents of the Nine-Point Circle at the midpoints of the opposite sides.

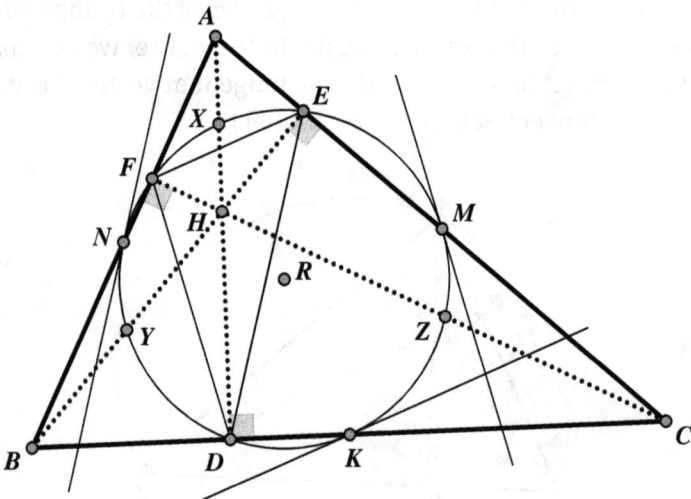

Figure 47

## Curiosity 48. An Adventure with the Inscribed Circle of a Triangle and a Special Circumcircle

In Figure 48, the randomly drawn triangle $ABC$ has point $I$ as the center of its inscribed circle. The circumscribed circle of triangle $BIC$ intersects $AB$ and $AC$ (extended) at points $D$ and $E$, respectively. Unexpectedly, we find that $AB = AE$ and that $AC = AD$.

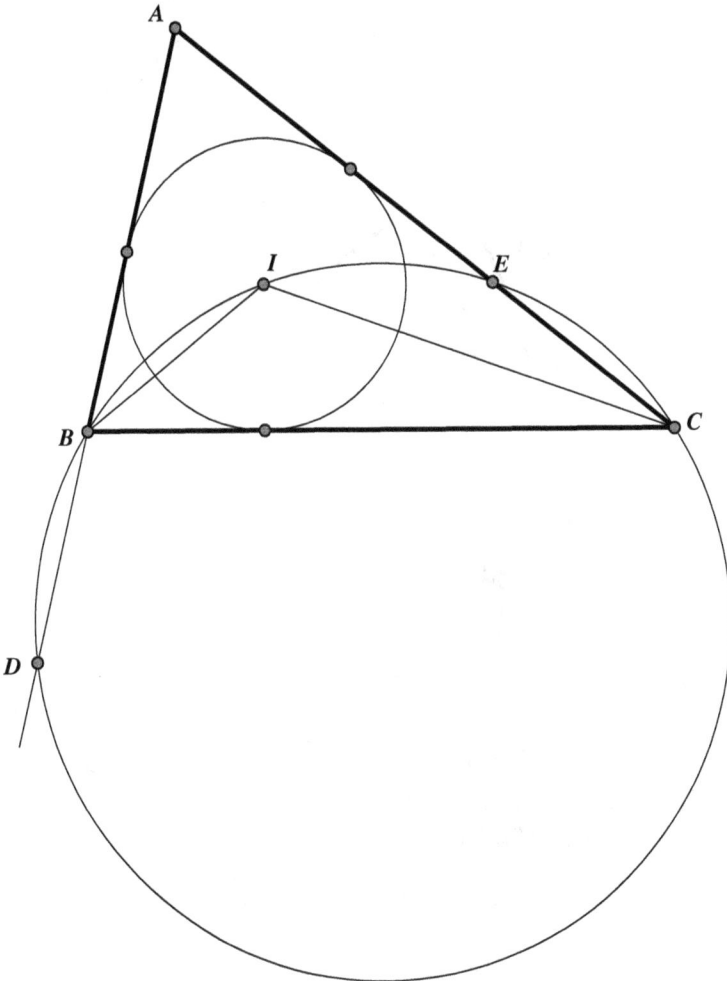

Figure 48

## Curiosity 49. Adventures with Circumscribed and Inscribed Circles of a Triangle

In Figure 49, point *I* is the incenter of triangle *ABC*. Points *P*, *Q*, and *R* are the circumcenters of triangles *BCI*, *CAI*, and *ABI*, respectively. Surprisingly, we find that *P*, *Q*, and *R* all lie on the circumscribed circle of triangle *ABC*.

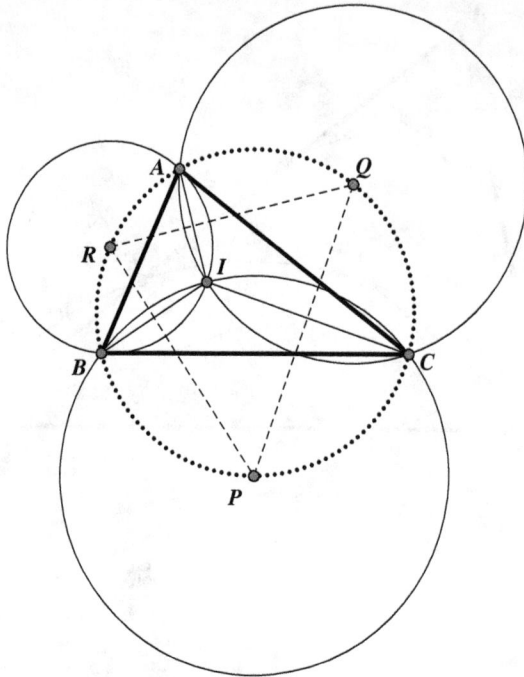

Figure 49

## Curiosity 50. More Adventures with Circumscribed and Inscribed Circles of a Triangle

The configuration in Curiosity 49 yields yet another amazing aspect. In Figure 50, point *I* is once again the incenter of triangle *ABC* and points *P*, *Q*, and *R* are again the circumcenters of triangles *BCI*, *CAI*, and *ABI*, respectively. It turns out that point *I* is the orthocenter (that is, the point of intersection of the altitudes) of triangle *PRQ*.

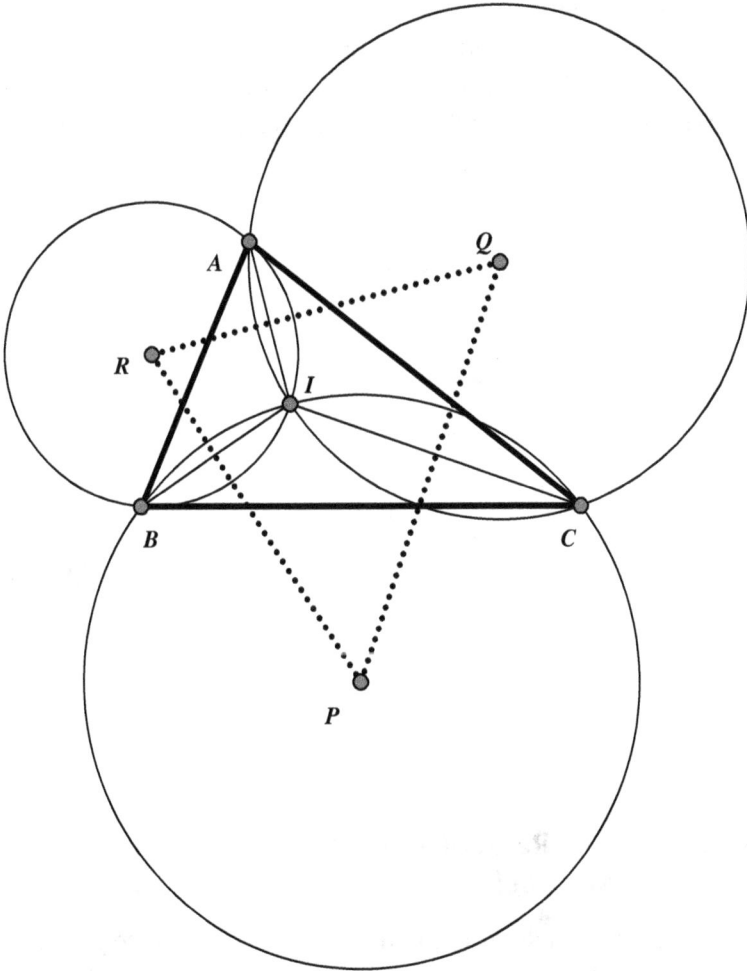

Figure 50

## Curiosity 51. How Four Centers of a Triangle can Generate an Unusual Cyclic Quadrilateral

In Figure 51, triangle *ABC* has a vertex angle $\angle BAC = 60°$. Point *H* is the orthocenter of triangle *ABC*, point *O* is its circumcenter, point *I* is its incenter, and point *D* is the center of its escribed circle (a circle tangent to two external sides and one internal side of the triangle) tangent to side *BC*. Unexpectedly, quadrilateral *DOIH* is a cyclic quadrilateral with *DO* = *DH*, and points *B* and *C* on its circumscribed circle *c*.

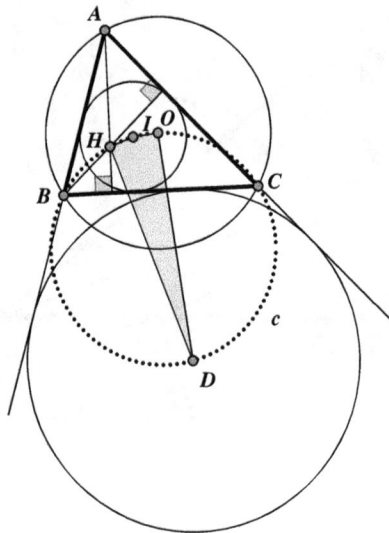

Figure 51

## Curiosity 52. The Radii of an Inscribed and Escribed Circle Determine a Rectangle Area

In Figure 52, in triangle *ABC*, the inscribed circle with center *I* is tangent to side *BC* at point *X*. The escribed circle with center *D* is tangent to side *BC* at point *Y*. An interesting relationship results, namely, that $CX \cdot BX = CY \cdot BY$, which also equals the area of a rectangle whose sides are the radii of the two circles.

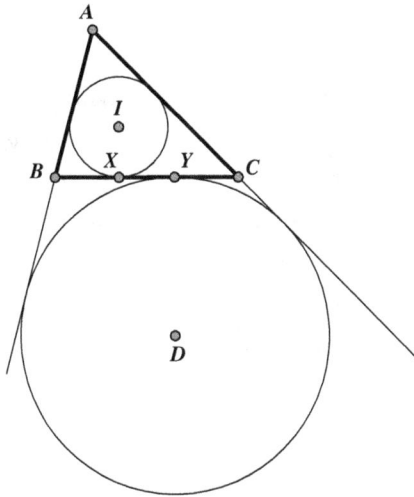

Figure 52

## Curiosity 53. A Circumscribed Circle and an Inscribed Circle Generate Three Equal-Length Lines

In Figure 53, point *I* is the center of the circle inscribed in triangle *ABC*, and point *O* is the center of the circumscribed circle of triangle *ABC*. When *BI* is extended to meet the circumscribed circle at point *P*, we determine three line segments of equal length: $PA - PC = PI$.

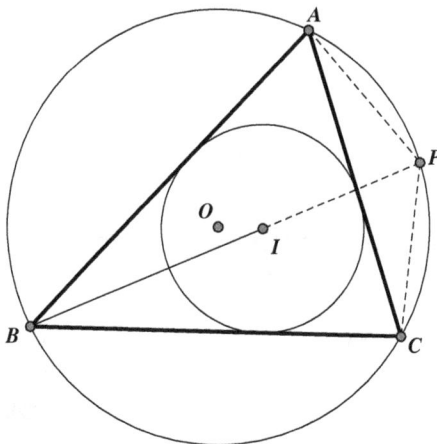

Figure 53

## Curiosity 54. Three Circles with a Surprising Common Point

In Figure 54, points $A$, $B$, $C$, $D$, $E$ and $F$ are placed on circle $c$, with center $O$, such that $AB = AF$, $CD = CB$, and $EF = ED$. When we draw the circle with center $A$ containing points $F$ and $B$, the circle with center $C$ containing points $B$ and $D$, and the circle with center $E$ containing points $D$ and $F$, unexpectedly, these three circles have a common point $H$.

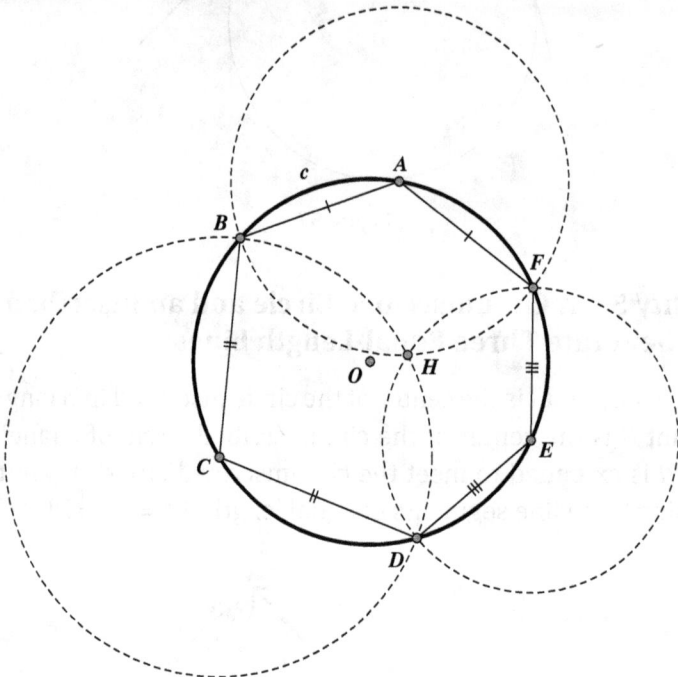

Figure 54

## Curiosity 55. The Famous Conway Circle

The sides of a triangle $ABC$ are extended, as shown in Figure 55, with $AD = AE = BC$, $BF = BG = CA$, and $CH = CJ = AB$. Surprisingly, points $D$, $E$, $F$, $G$, $H$, and $J$ all lie on a common circle, whose center $I$ is the center of the inscribed circle of triangle $ABC$. This circle is commonly referred to as the *Conway circle*, after the English mathematician John H. Conway (1937–2020).

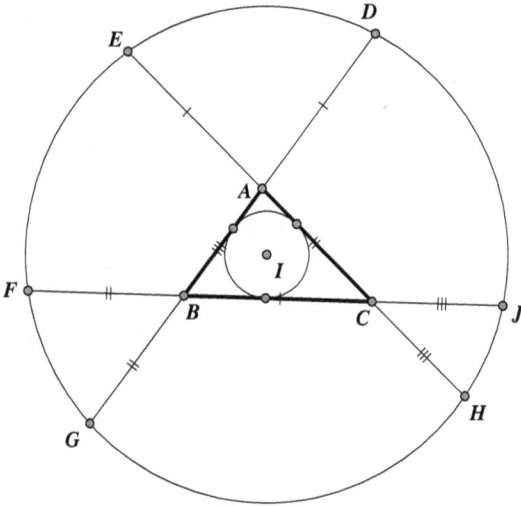

Figure 55

## Curiosity 56. A Surprising Collinearity with the Incenter of a Triangle — Sawayama's Lemma

We are given a triangle *ABC* and a point *P* on side *BC*, as shown in Figure 56. The circle with center *M* is tangent to the circumcircle of *ABC* at a point *D*, tangent to side *BC* at point *E*, and tangent to line *AP* at point *F*. We find that points *E* and *F* are collinear with the incenter *I* of triangle *ABC*. This interesting result is commonly known as *Sawayama's Lemma*.

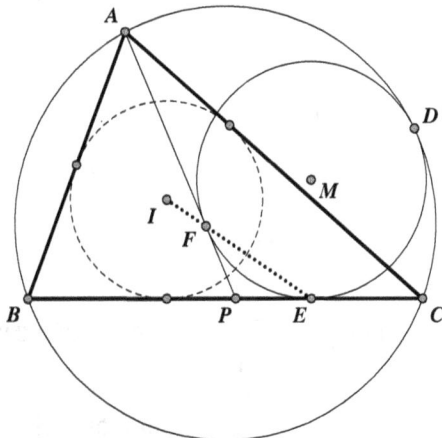

Figure 56

## Curiosity 57. The Surprising Relationship Between the Four Tangent Circles of a Triangle

In Figure 57, triangle *ABC* has an inscribed circle and three escribed circles. Point *I* is the incenter of *ABC*, while points *D*, *E*, and *F* are the centers of the escribed circles on sides *AB*, *BC*, and *CA*, respectively. The square of the distance between the centers of two of these circles is equal to the square of the distance between the centers of the two other circles. Furthermore, this is also equal to the square of the diameter of the circle through the centers *D*, *E*, and *F*. In Figure 57, this relationship is seen as $DE^2 + FI^2 = DF^2 + EI^2 = DN^2$, where *DN* is a diameter of circle *M*, which contains points *D*, *E*, and *F*.

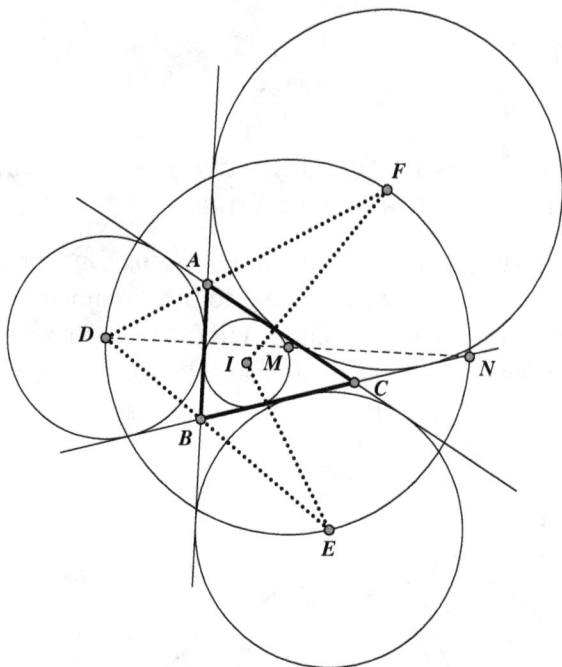

Figure 57

## Curiosity 58. Two Tangent Circles Generate an Unexpected Right Angle

Circles with centers *O* and *Q* are tangent at point *P*, as shown in Figure 58. The common tangent line *RS* is tangent to circle *O* at *R* and to circle $Q$ at $S$. A randomly-chosen line through point $P$ intersects the two circles $O$ and $Q$ at points $A$ and $B$, respectively. Unexpectedly, when $AR$ extended meets $BS$ extended at point $H$, the angle $\angle AHB$ is a right angle.

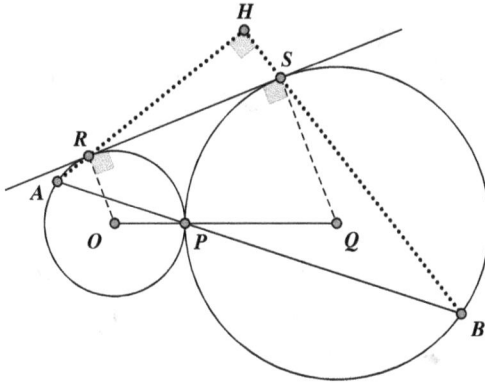

Figure 58

## Curiosity 59. The Unexpected Diameter of a Circle

The diameter of two tangent equal circles arises in a very unexpected fashion. In Figure 59, equal circles with centers $O$ and $Q$ are tangent at point $P$. A chord $PB$ is randomly drawn in circle $Q$. The line $AP$ is drawn perpendicular to $PB$ at point $P$ and intersects circle $O$ at point $A$. Quite unexpectedly, we find that $AB$ is equal to the sum of the radii of the two circles, so that $AB = OQ$.

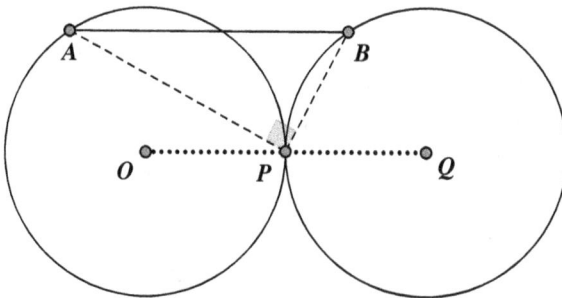

Figure 59

## Curiosity 60. Doubling the Distance Between Centers

Two circles, with centers $O$ and $Q$, intersect at points $A$ and $B$, as shown in Figure 60. The line through point $A$ and parallel to $OQ$ intersects circle $O$ at point $X$ and circle $Q$ at point $Y$, with the curious result that $XY = 2 \cdot OQ$.

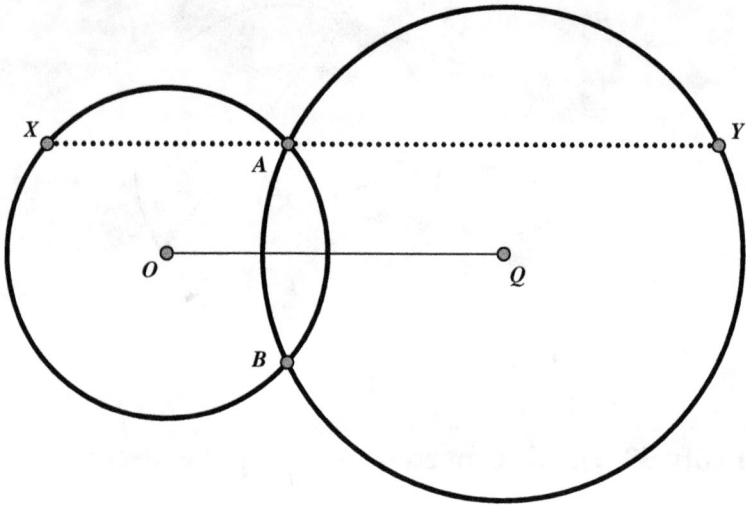

Figure 60

## Curiosity 61. Tangent Circles Generating a Circle Tangent to the Line of Centers

When the common external tangent of two externally tangent circles is the diameter of a third circle, that third circle is tangent to the line joining the centers of the initial two circles at their common point. We see this demonstrated in Figure 61, where $AB$ is the common external tangent to the two tangent circles with centers $O$ and $Q$, which are tangent at point $P$. Quite unexpectedly, the circle with diameter $AB$ is tangent to line $OQ$ at point $P$.

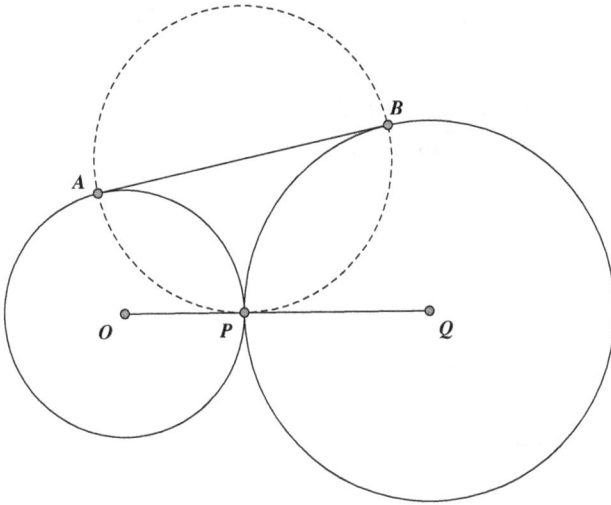

Figure 61

## Curiosity 62. An Unexpected Angle Bisector Generated by Externally Tangent Circles

In Figure 62, externally tangent circles with centers $O$ and $Q$ share point $P$. The tangent line to circle $Q$ at a random point $A$ intersects circle $O$ at points $X$ and $Y$, and line $YP$ extended intersects circle $Q$ a second time at point $B$. Surprisingly, we find that line $PA$ is the bisector of angle $\angle XPB$.

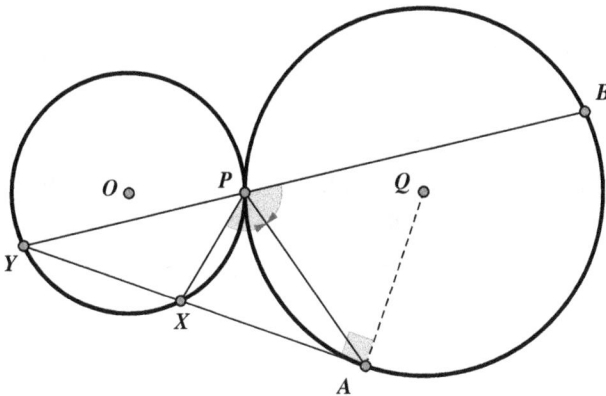

Figure 62

### Curiosity 63. Unexpected Parallel Chords in Externally Tangent Circles

In Figure 63, we are given circles with centers $O$ and $Q$, which are externally tangent at point $P$. The tangent line to circle $Q$ at a random point $A$ intersects circle $O$ at points $X$ and $Y$. Line $YP$ extended intersects circle $Q$ a second time at point $B$, and line $AP$ extended intersects circle $O$ a second time at point $C$. Quite surprisingly, we find that lines $AB$ and $YC$ are parallel.

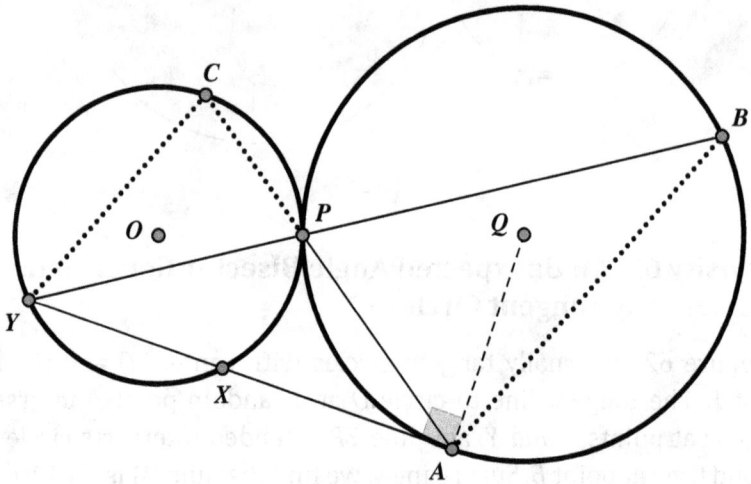

Figure 63

### Curiosity 64. Surprising Parallel Tangents of Externally Tangent Circles

In Figure 64, we are given two circles, externally tangent at point $P$, with centers $O$ and $Q$. A random line containing point $P$ intersects circle $O$ a second time at point $A$ and circle $Q$ a second time at point $B$. Amusingly, we find that the tangent lines at points $A$ and $B$ are parallel.

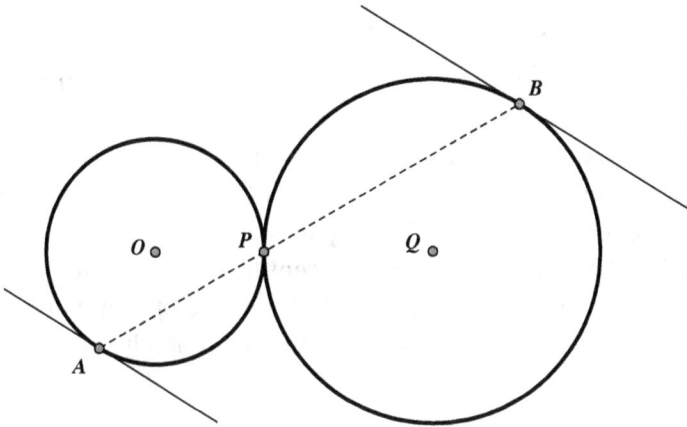

Figure 64

## Curiosity 65. Two Externally Tangent Circles Generate a Line Half as Long as Their Line of Centers

Two externally tangent circles with centers $O$ and $Q$ share point $A$, as shown in Figure 65. Points $B$ and $C$ are chosen on circles $O$ and $Q$, respectively, so that line $BC$ is a common tangent of the circles. The line perpendicular to $BC$ through $A$ and the perpendicular bisector of $OQ$ intersect at point $P$. We then find that $OQ = 2AP$.

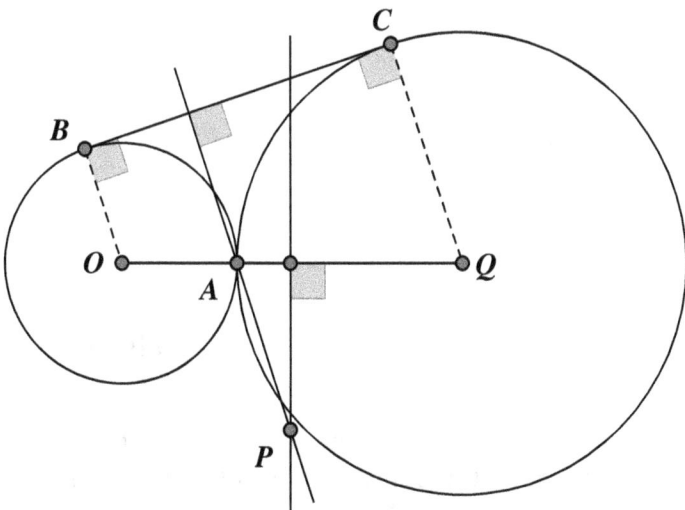

Figure 65

### Curiosity 66. A Very Surprising Constant Tangent Length

Among the many astounding relationships we have encountered, this Curiosity is exceptionally unusual. In Figure 66, we are given a circle with center $O$, which is intersected at points $A$ and $B$ by a secant. Point $M$ is the midpoint of arc $AB$ on circle $O$. We draw any circle $Q$ on the same side of line $AB$ as point $M$, tangent to circle $O$ externally, and tangent to line $AB$. The length of the tangent from $M$ to circle $Q$ is always of a constant length $t$. In other words, the value of $t$ only depends on the choice of circle $O$ and secant $AB$, and is completely independent of the choice of circle $Q$.

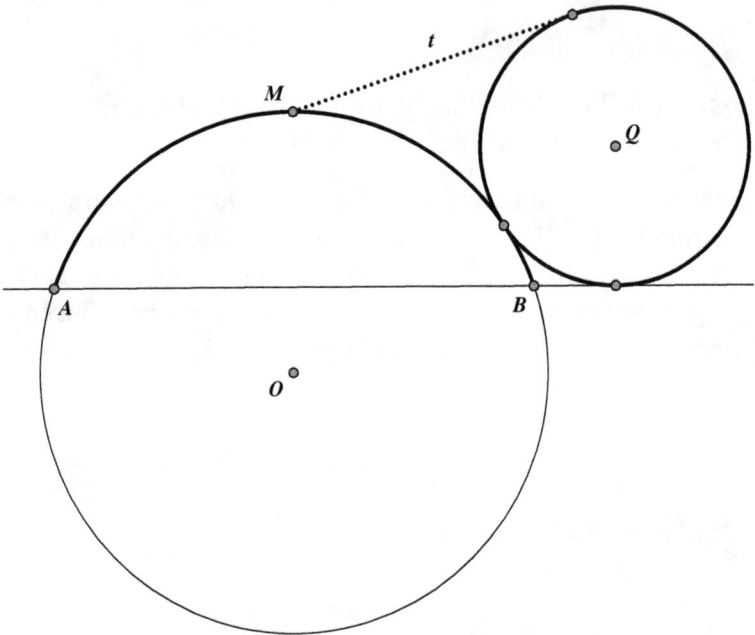

Figure 66

### Curiosity 67. Two Circles Creating Parallel Lines Unexpectedly

When we draw two intersecting lines through the point of tangency of two tangent circles, the points in which these lines intersect the circles determine a pair of parallel lines. In Figure 67, lines *APB* and *DPC* contain the point of tangency *P* of the two circles with centers *O* and *Q*. The points where these two lines intersect the circles then determine parallel lines *AD* and *BC*.

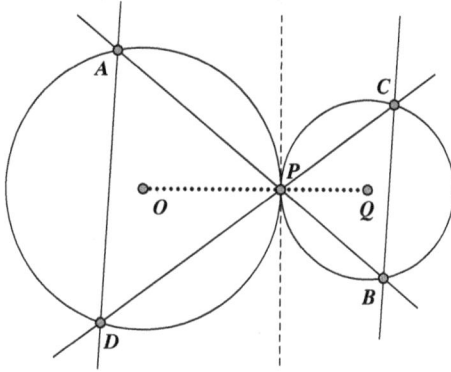

Figure 67

## Curiosity 68. Two Circles Create an Angle Bisector

In Figure 68, the two circles with centers *O* and *Q* are tangent at point *A*. Chord *BC* of circle *O* is tangent to circle *Q* at point *D*. Quite unexpectedly, we find that $\angle BAD = \angle DAC$.

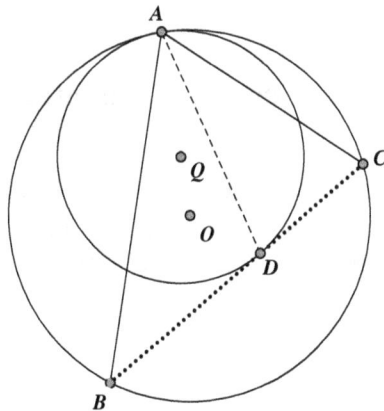

Figure 68

## Curiosity 69. Tangent Circles Produce Unusual Line Products

In Figure 69, circles with centers $O$ and $Q$ are internally tangent at point $C$. Chord $AB$ of circle $O$ is tangent to circle $Q$ at point $D$, and $CD$ extended intersects circle $O$ a second time at point $G$. We then find that $AC \cdot BC = CD \cdot CG$.

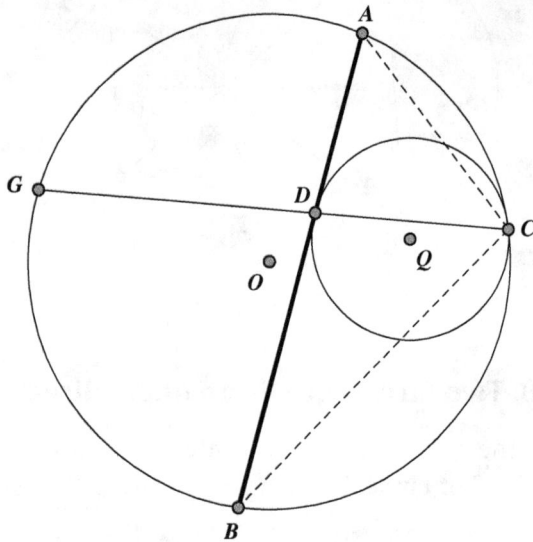

Figure 69

## Curiosity 70. Equal Angles Appear in Internally Tangent Circles

In Figure 70, two internally tangent circles with centers $O$ and $Q$ share the common tangency point $P$. A random line intersects the larger circle at points $A$ and $B$ and the smaller at points $C$ and $D$. Independent of our choice of line, we find that $\angle CPA = \angle BPD$.

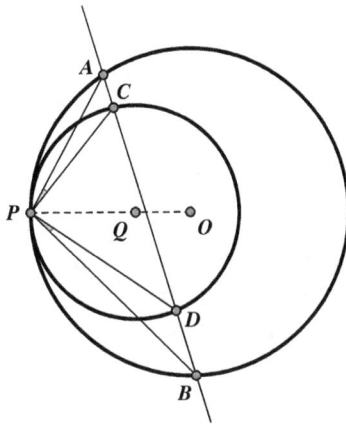

Figure 70

## Curiosity 71. Incredible Tangent Lengths in Tangent Circles

In Figure 71, we are given circles with centers $O$ and $Q$ that touch internally at point $P$. The vertices $A$, $B$, and $C$ of an equilateral triangle are chosen at random on the larger of the two circles. When we draw tangents from the vertices of this equilateral triangle to the smaller circle, we are surprised to find that the sum of the lengths of the two shorter tangents is equal to the length of the longer tangent. In this particular case, we have $BE = AD + CF$.

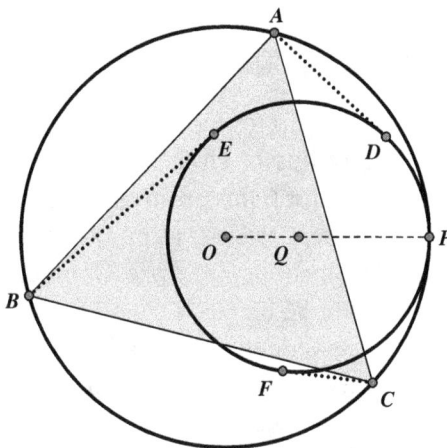

Figure 71

## Curiosity 72. Symmetric Equal Angles

In Figure 72, the two circles with centers $O$ and $Q$ are internally tangent at point $A$. Point $P$ is randomly selected, and does not lie on the line $OQ$, nor does it lie on either of the circles. Points $E$ and $R$ lie on circle $O$ such that $E$ is the point of the circle situated furthest from $P$, and $R$ is the point of the circle situated nearest to $P$. Similarly, points $F$ and $S$ lie on circle $Q$ such that $F$ is the point of the circle situated furthest from $P$, and $S$ is the point of the circle situated nearest to $P$. We are then amused to find that $\angle EAF = \angle RAS$.

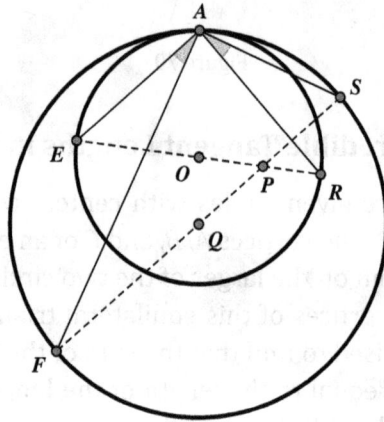

Figure 72

## Curiosity 73. An Amazing Rolling Result

On the left side of Figure 73, circles with centers $O$ and $Q$ are internally tangent at point $A$. The radius of circle $O$ is twice the length of the radius of circle $Q$, so that point $O$ lies on circle $Q$. We draw radii $OB$ and $OC$ of circle $O$, which intersect circle $Q$ at points $D$ and $E$, respectively. We then find that arc $AB$ on circle $O$ and arc $AD$ on circle $Q$ are equally long. This is also true of arc $BC$ on circle $O$ and arc $DE$ on circle $Q$.

A truly astounding consequence of this is illustrated on the right side of Figure 73. We roll circle $Q$ without slipping along the interior of circle $O$ until points $B$ and $D$ coincide, and find that the point $X$ on circle $Q$, which coincided with point $O$ in the original position, comes to lie on the diameter of circle $O$ perpendicular to $OA$. In fact, if circle

$Q$ completes the rotation around the interior of circle $O$, point $X$ will traverse this entire diameter of circle $O$.

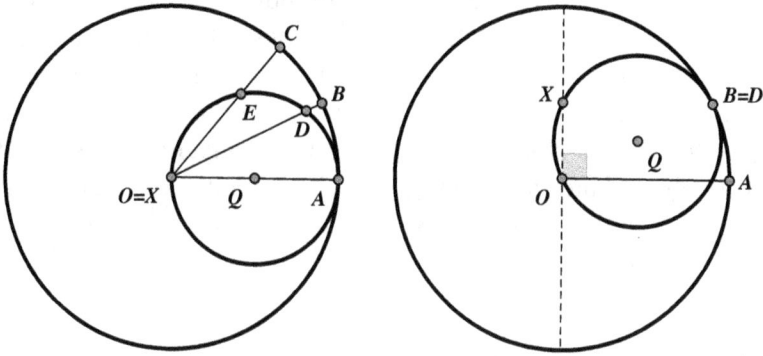

Figure 73

## Curiosity 74. Semicircles — The Arbelos

The semicircle also possesses some fascinating properties. An example of this is the *arbelos*, which is sometimes referred to as the *shoemaker's knife*. This interesting figure, shown as the lighter shaded region in Figure 74, is formed by three tangent semicircles with diameters on a common line. This means that the sum of the diameters of the two smaller semicircles is equal to that of the larger semicircle. Although it may not be intuitively obvious, the sum of the lengths of the two smaller semicircular arcs is equal to the length of the larger semicircular arc.

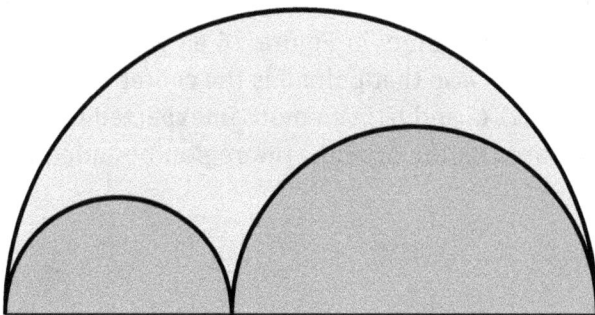

Figure 74

## Curiosity 75. An Arbelos Bisection

In Figure 75, we add a few lines to the arbelos, which result in an interesting feature. Points $A$ and $B$ are the endpoints of the large semicircular arc, and point $C$ is the tangent point of the two smaller arcs. We construct a perpendicular line to the line segment $AB$ at point $C$ to intersect the larger semicircle at point $H$. We then draw a common tangent to the two smaller semicircles, with tangent points $F$ and $G$, respectively, and intersecting segment $HC$ at point $S$. We then find that the two segments $CH$ and $FG$ bisect each other at point $S$, so that $SF = SG$ and $SC = SH$.

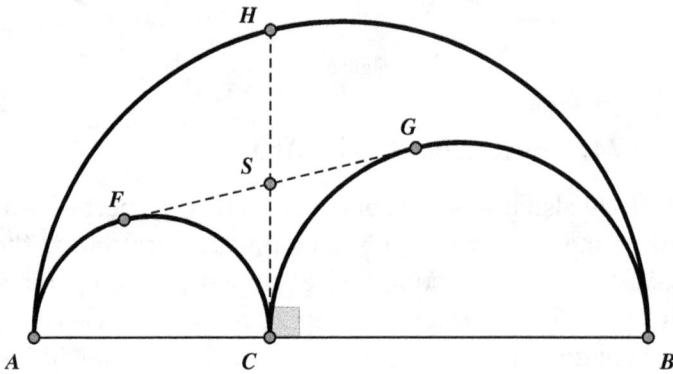

Figure 75

## Curiosity 76. More on the Common Arbelos Bisection Point

The configuration of Curiosity 75 yields even more wonderful results. Defining the points in Figure 76 as we did in Curiosity 75, we find that $HC = FG$, so that point $S$ is the center of a circle containing the points $F$, $C$, $G$, and $H$. Even more unexpected is that this circle has the same area as the arbelos, the region bounded by the three semicircles.

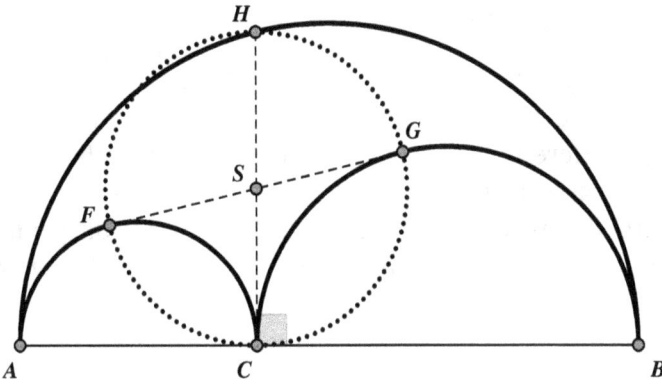

Figure 76

## Curiosity 77. Yet More on the Arbelos

Continuing our exploration of the arbelos in Figure 77, there is another unexpected geometric wonder to be discovered. Once again naming all points as in Curiosity 75, we find that line segments $AH$ and $BH$ contain the tangency points of the common tangent line $F$ and $G$, respectively, giving us the further surprising collinearities $AFH$ and $BGH$.

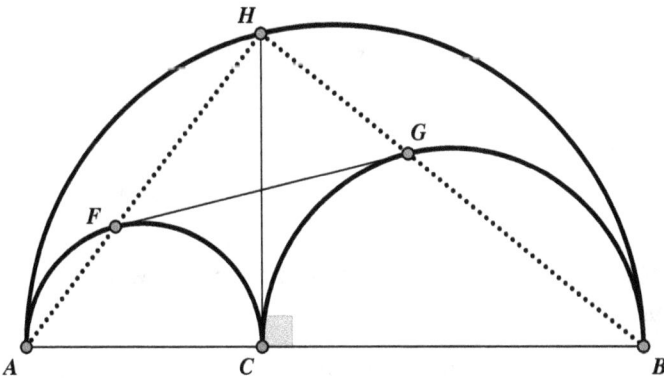

Figure 77

## Curiosity 78. Another Arbelos Surprise

There are many surprise relationships relating to the arbelos. Another is shown in Figure 78, where points $A$, $B$, and $C$ are again defined as in Curiosity 75. Here, points $S$ and $T$ are the midpoints of the two smaller semicircular arcs, and point $R$ is the midpoint of the reflection in $AB$ of the larger semicircular arc. Amazingly, the area of quadrilateral $CSRT$ is equal to the sum of the squares of the radii of the two smaller semicircles.

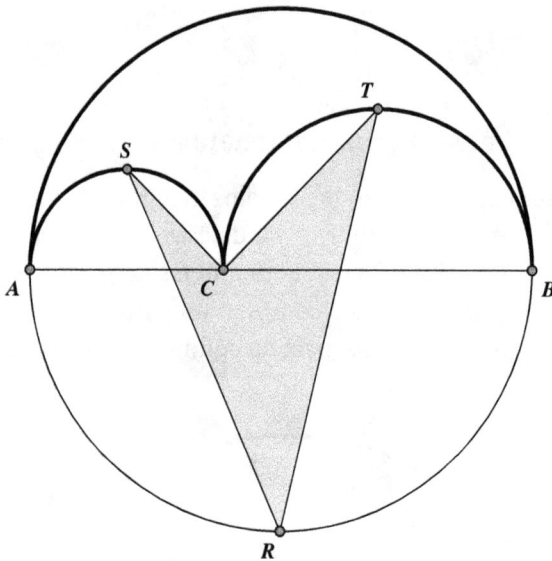

Figure 78

## Curiosity 79. Some Further Entertainment with the Arbelos

Curiosity 76 presented a circle with the same area as the arbelos. Surprisingly, we can find another circle with a similar property. In Figure 79, points $A$, $B$, and $C$ are again as defined in Curiosity 75. The arc $AC$ of the arbelos is reflected along $AC$, resulting in a shape whose edge is composed of three semicircles $AB$, $BC$, and the lower arc $AC$, as shown. A tangent from point $A$ is then drawn to the semicircle $BC$ at point $T$. Finally, a circle is constructed with center $R$, with $AT$ as its diameter. The unexpected result is that the area of the shape bordered by the three arcs is equal to the area of the circle with center *R*.

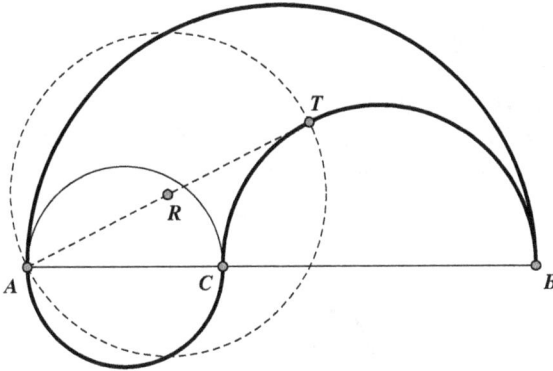

Figure 79

## Curiosity 80. A Surprising Area Equality Created by Semicircles

Consider right triangle *ABC* with semicircles on each of its sides, as shown in Figure 80. Point *D* on side *AB* is the intersection of the two semicircles on sides *AC* and *BC*. Line *CD* is then added. The areas of the bordered regions *x*, *y*, and *s*, as shown in Figure 80, enable us to establish that $s - x - y = $ area$[ABC]$.

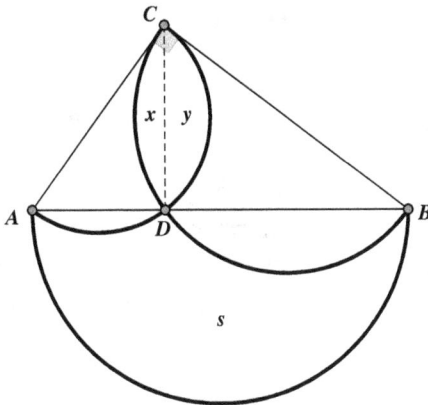

Figure 80

## Curiosity 81. Another Surprising Area Equality Created by Semicircles

In Figure 81, we are presented with a similar situation to Curiosity 80. Once again, a right triangle *ABC* is drawn with semicircles on the sides, as shown. As before, point *D* is the intersection of the semicircles on the side *AB* of triangle *ABC*, and line *CD* is added. Then, a circle with diameter *CD* is drawn, as are semicircles with diameters *AD* and *BD*. These circles border regions *j*, *k*, *p*, and *q*, as shown in Figure 81. Amazingly, we find that $j + k + p + q = $ area[*ABC*].

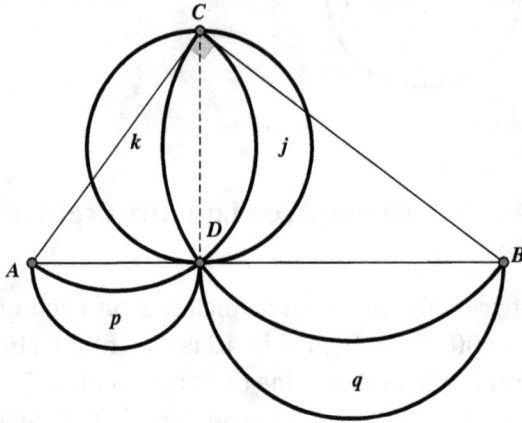

Figure 81

## Curiosity 82. An Amazing Curve Division into Two Equal Parts

We are given a semicircle with diameter *AB*, as shown in Figure 82. Point *M* is the midpoint of *AB*. When we add semicircles with diameters *AM* and *MB* external to the large semicircle, we obtain a closed curve composed of three semicircles. Amazingly, if we draw any line through point *M*, intersecting the closed curve at points *P* and *Q*, these points divide the closed curve into two sections of equal length.

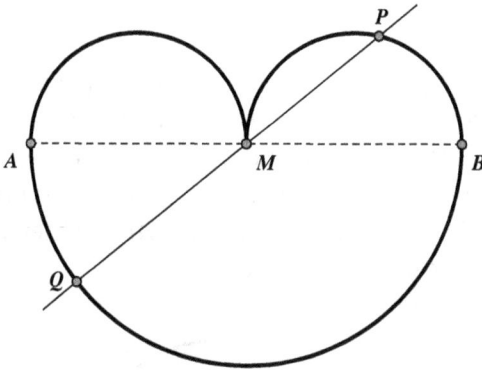

Figure 82

## Curiosity 83. An Unexpected Bisection

In Figure 83, beginning with segment *AB* with midpoint *M*, circular arcs with diameters *AB* and *MB* are drawn on the same side of line *AB*. Points *P*, *Q* and *R* are selected on the smaller of these semicircles, so that arcs $\overset{\frown}{BR}, \overset{\frown}{RQ}$ and $\overset{\frown}{PQ}$ are equal. Line *MQ* intersects the larger semicircle at point *C* and *MQ* intersects line *PB* at point *D*. Surprisingly, we find that *Q* is the midpoint of *CD*.

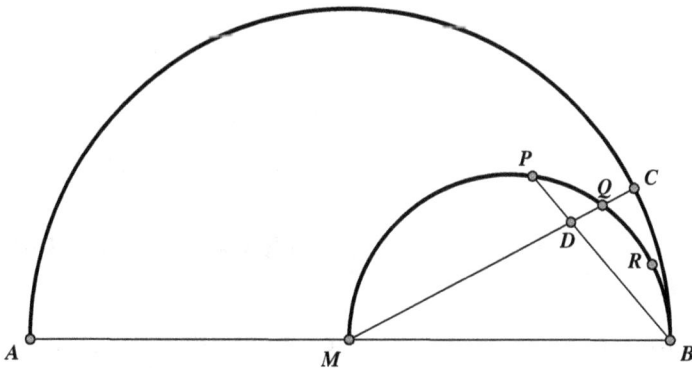

Figure 83

## Curiosity 84. A Circle-Related Point with a Double Function

Three mutually tangent circles generate a special point that is simultaneously the circumcenter of one triangle and the incenter of another. We see this illustrated in Figure 84, where circles with centers $R$, $S$, and $T$ are tangent at points $A$, $B$, and $C$, forming two triangles: $\triangle RST$ and $\triangle ABC$. The circumcenter $P$ of triangle $ABC$ is then also the center of the inscribed circle of triangle $RST$.

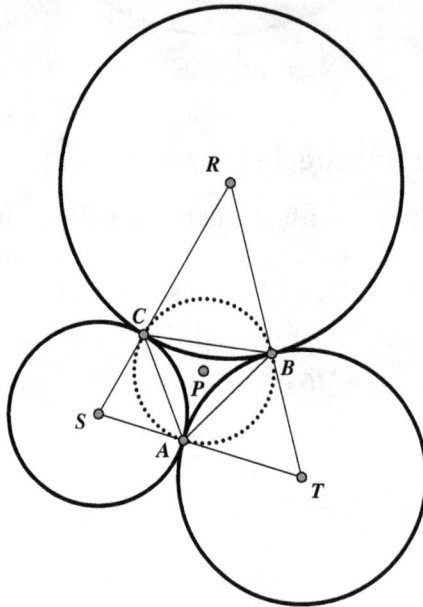

Figure 84

## Curiosity 85. Three Circles Through Two Common Points Generate a Surprising Circle

When three circles share two common points, their tangents can generate a fourth circle. In Figure 85, three circles with centers $X$, $Y$, and $Z$, respectively, share the common points $A$ and $B$. A line through point

*A* intersects these three circles at points *D*, *E*, and *F*, respectively. The tangents at points *D* and *E* meet at point *P*, the tangents at points *E* and *F* meet at point *Q*, and the tangents at points *D* and *F* meet at point *R*. Quite unexpectedly, we find that the points *P*, *B*, *Q*, and *R* all lie on the same circle, thus making quadrilateral *PBQR* cyclic.

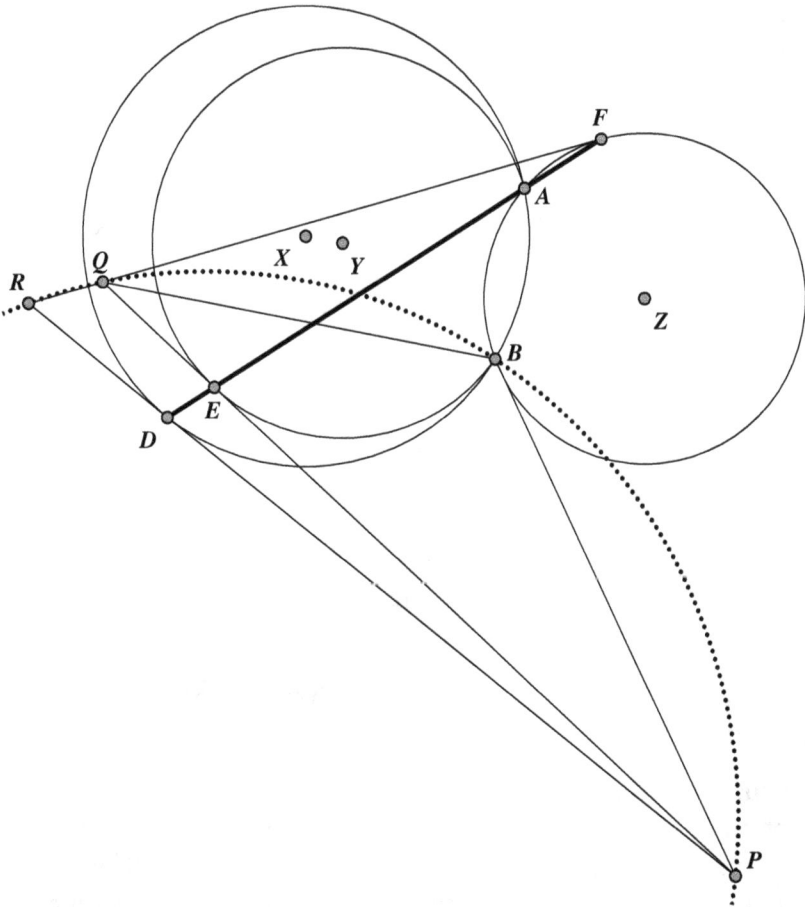

Figure 85

## Curiosity 86. Intersecting Circles Create Five Concyclic Points

Getting more than four points on one circle is always delightful. Consider circles $O$ and $Q$ intersecting at points $P$ and $R$, as shown in Figure 86. When we extend $PQ$ to intersect circle $O$ at point $A$, and $OP$ to intersect circle $Q$ at point $B$, we find that points $A$, $O$, $R$, $Q$, and $B$ are all on the same circle.

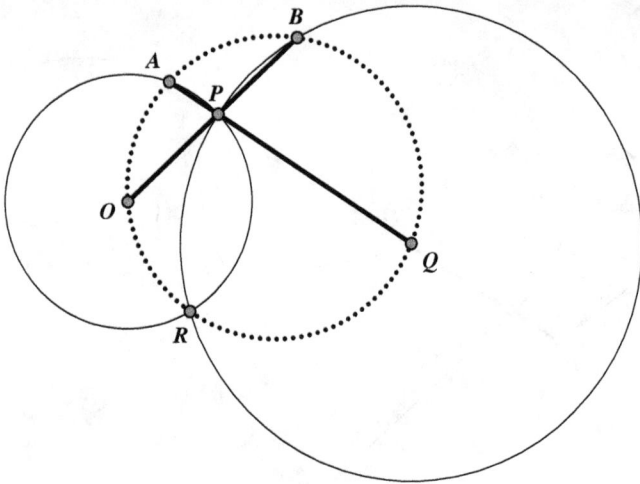

Figure 86

## Curiosity 87. Concyclic Centers of Three Related Circumcircles

In Figure 87, we are given three collinear points $A$, $B$, and $C$, and a point $P$ not on their common line. Points $R$, $S$, and $T$ are the circumcenters of triangles $PAB$, $PBC$, and $PAC$, respectively. In this relatively simple configuration, we find that the three circumcenters are concyclic with the point $P$.

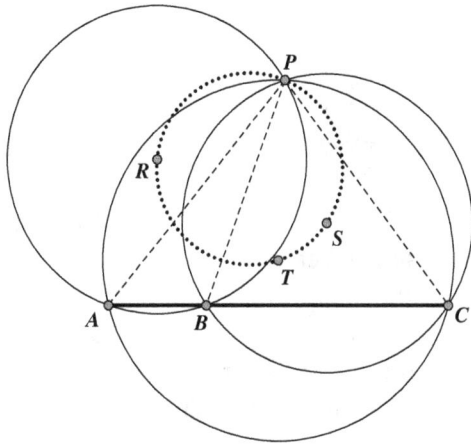

Figure 87

## Curiosity 88. Two Unequal Circles Counterintuitively Produce Two Equal Circles

We begin with two unequal circles, with centers at $O$ and $Q$, that intersect at points $P$ and $R$. A common tangent touches the circles at points $A$ and $B$, respectively. Quite surprisingly, the circumcircles of triangles $APB$ and $ARB$ are of equal size, as we see in Figure 88.

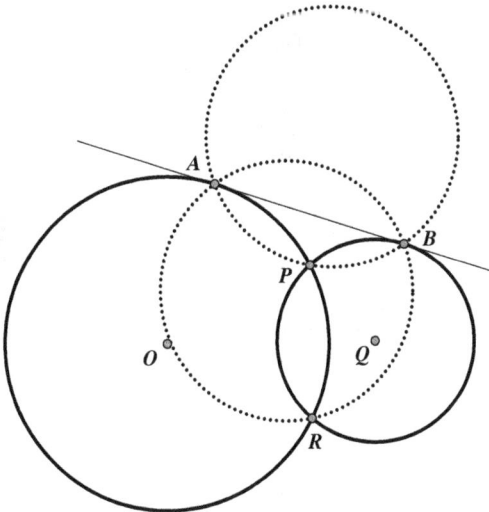

Figure 88

## Curiosity 89. An Angle Property of the Miquel Point of a Triangle

If we are given a triangle *ABC* and choose points *D*, *E*, and *F* on the lines *BC*, *CA*, and *AB*, respectively, the circumcircles of triangles *AFE*, *BDF*, and *CED* contain a common point *M*. This point is known as the *Miquel point*. This property is independent of the positioning of the points. In Figure 89a, all points have been chosen on the triangle sides, while in Figure 89b, only point *E* is on a triangle side, while *D* and *F* are chosen on the extensions of sides *BC* and *AB*. In both cases, each of the circles contains two points on the sides of triangle *ABC* (or their extensions) and the vertex between these sides. This leads to the unanticipated fact that the three circles will always contain a common point *M*.

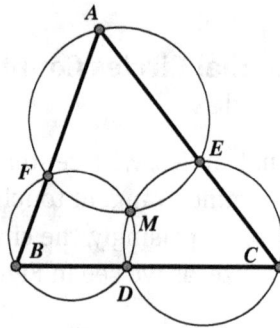

Figure 89a

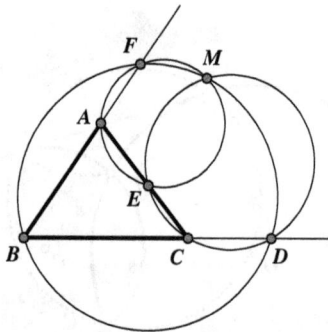

Figure 89b

The segments joining the Miquel point *M* of a triangle to the points *D*, *E*, and *F* chosen on the triangle sides form congruent angles with the respective sides of the triangle. In Figure 89c, we have ∠*AEM* = ∠*BFM* = ∠*CDM*, and in Figure 89d, where we have chosen point *X* on the extension of *BA*, we have ∠*MEA* = ∠*MDB* = ∠*MFX*.

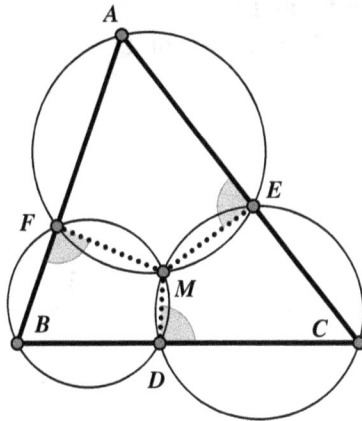

Figure 89c

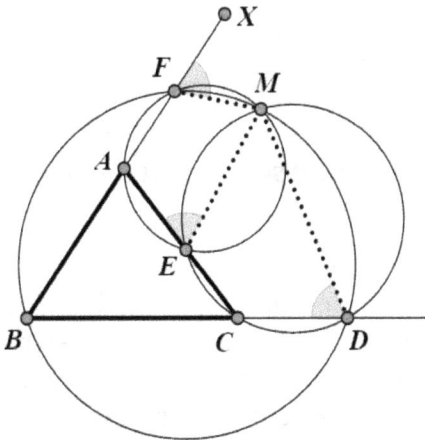

Figure 89d

## Curiosity 90. Yet More on the Miquel Relationship

In Figure 90, we are given four lines that intersect in six points *A*, *B*, *C*, *D*, *E*, and *F*. These lines form the four triangles *ADB*, *BEF*, *CDE*, and *ACF*. When we draw the circumscribed circles about each of these triangles, unexpectedly, we find that they all contain a common point, *M*, which we name their *Miquel point*. Furthermore, the centers of the four circles, *P*, *Q*, *R*, and *S*, all lie on the same circle along with point *M*, yielding five unexpected concyclic points.

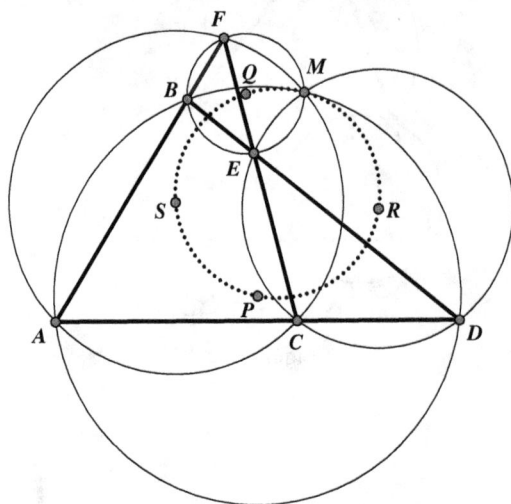

Figure 90

## Curiosity 91. Three Intersecting Circles Generate Similar Triangles and More

The three intersecting circles shown in Figure 91 share a common point *O*. Triangle *ABC* is formed by joining the centers of these three circles. The lines joining point *O* with each of the circles' centers intersect the respective circles at points *D*, *E*, and *F*. It then turns out that the triangles *ABC* and *DEF* have parallel sides. Furthermore, we find that the sides of triangle *DEF* contain the intersection points *P*, *Q*, and *R* of the pairs of circles, as shown.

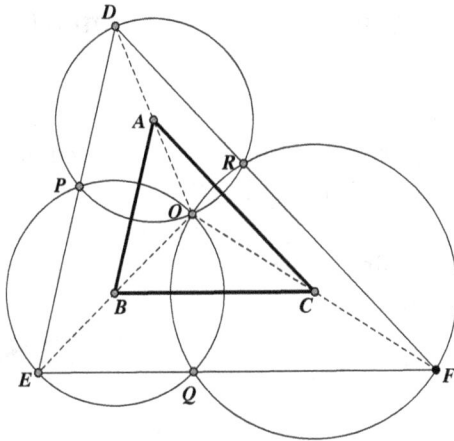

Figure 91

## Curiosity 92. Equal Circles Generated by Three Equal Circles with a Common Point

When three circles of equal size pass through a common point $P$, as shown in Figure 92, the circle through their centers is of equal size to the original three circles. In the figure, we see circles with centers $Q$, $R$, and $S$, and the common point $P$. Surprisingly, we find that the other points $A$, $B$, and $C$ in which these circles intersect determine another circle of the same size.

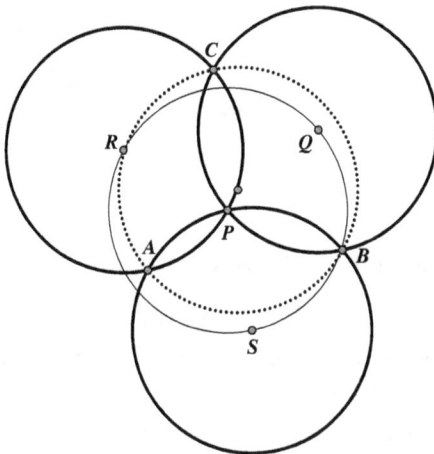

Figure 92

## Curiosity 93. Congruent Figures Generated by Three Equal Circles with a Common Point

There is another interesting aspect to the configuration in Curiosity 92. As we see in Figure 93, if point $D$ is the circumcenter of triangle $ABC$, the quadrilaterals $APBC$ and $QDRS$ are congruent. Furthermore, in either of these congruent figures, the line joining any two points is perpendicular to the line joining the other two points. For instance, we have $AP \perp BC$ and $BP \perp CA$.

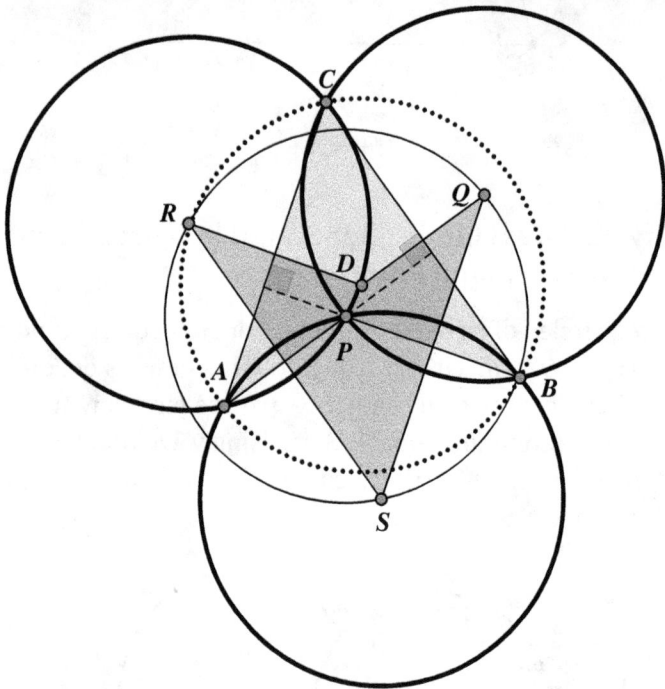

Figure 93

## Curiosity 94. Fun with Intersecting Circles

An amazing property results when we select four points $A$, $B$, $C$, and $D$ on a circle, as shown in Figure 94. When we draw circles containing each pair of consecutively placed points, and naming the other points at which pairs of these consecutive circles intersect as $K$, $L$, $M$, and $N$, respectively, we find that these points also lie on a common circle.

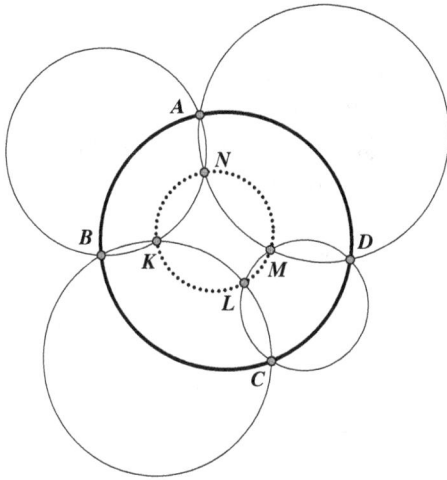

Figure 94

## Curiosity 95. Four Equal Circles Generate a Circumscribable Quadrilateral

In Figure 95, we are given four equal circles with centers $A$, $B$, $C$, and $D$. All four circles contain a common point $P$. The common tangents of these four circles create a quadrilateral $EFGH$, as shown. Amazingly, the quadrilateral $EFGH$ can be circumscribed; that is, the four points $E$, $F$, $G$, and $H$ are concyclic.

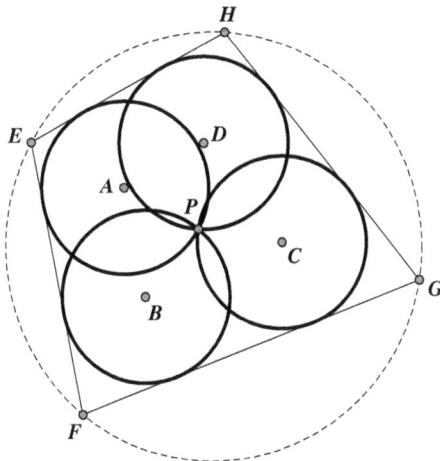

Figure 95

## Curiosity 96. The Reuleaux Triangle

In Figure 96, we consider the rather odd-looking shape called a *Reuleaux triangle*, named after the German engineer Franz Reuleaux (1829–1905). The Reuleaux triangle is formed by three circular arcs with a radius equal to the length of the sides of an equilateral triangle $ABC$ and centers in the vertices of $ABC$. Wherever we place a pair of parallel tangents, they will always be the same distance apart. We name this distance $b$ the breadth of the Reuleaux triangle. In the figure, this constant distance is illustrated by the segments $AP = CQ = b$. Surprisingly, the perimeter of the Reuleaux triangle with breadth $b$ is equal to the perimeter of a circle with diameter $b$.

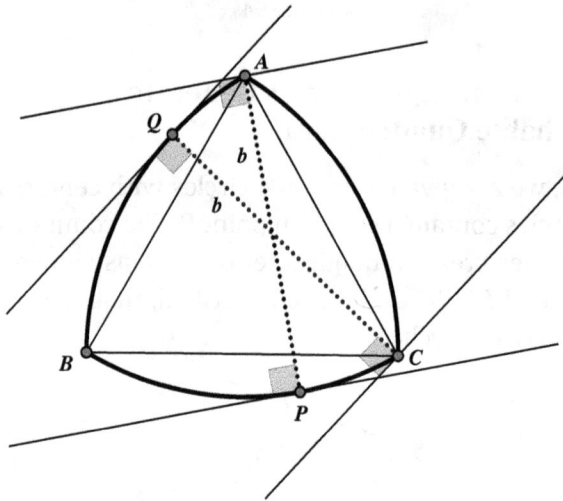

Figure 96

## Curiosity 97. Some Other Shapes with Constant Breadth

The Reuleaux triangle, which we were introduced to in Curiosity 96, and the circle are not the only shapes with constant breadth. In Figure 97, we see two more examples of such shapes. On the left, we start from a regular pentagon *ABCDE*, and consider the shape formed by the five circular arcs centered in the five vertices that contain the endpoints of their respective opposite sides. Similarly, on the right, we have a shape produced in the same way, but starting with a regular heptagon *FGHIJKL*. Both of these shapes also have constant breadth *b*, and a perimeter equal to that of a circle with diameter *b*. In fact, we can produce a shape with constant breadth using by drawing arcs in this way on any regular polygon with an odd number of sides.

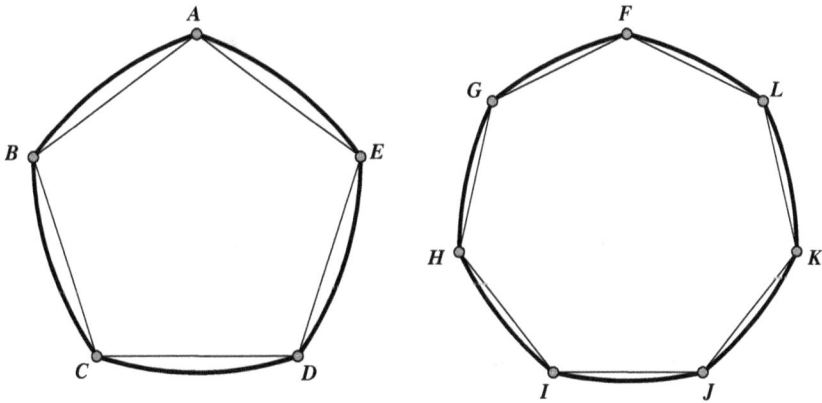

Figure 97

## Curiosity 98. Introducing the Radical Axis of Intersecting Circles

Consider two circles with centers $O$ and $Q$, as shown in Figure 98, which intersect at points $A$ and $B$. We wish to find points $P$ with the property that the tangents from $P$ to the two circles are of equal length. These points are exactly the points on the line $AB$, external to the circles. We call this line $AB$ the *radical axis* of the two circles. All points $T$ on the radical axis have the property that the equality $TU \cdot TV = TW \cdot TX$ holds for all secants $TUV$ of circle $O$ and all secants $TWX$ of circle $Q$.

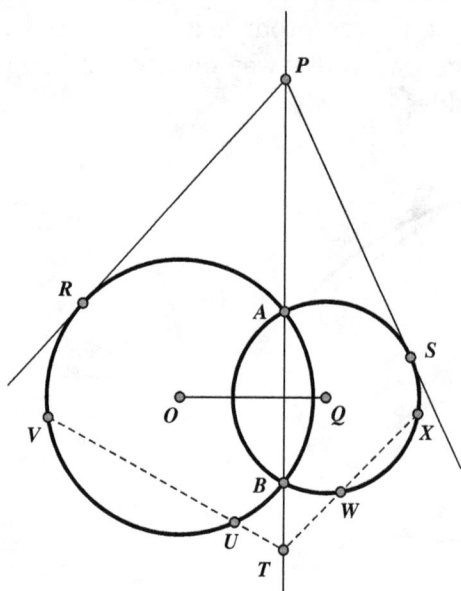

Figure 98

## Curiosity 99. The Radical Axis of Tangent Circles

In Figure 99, two circles with centers $O$ and $Q$ are tangent at point $Z$. The circles are externally tangent in Figure 99a and internally tangent in Figure 99b. In each case, the radical axis of the two circles is their common tangent at point $Z$. As in the intersecting case we considered in Curiosity 98, points $P$ on this line have the property that the tangents from $P$ to the two circles are of equal length. Also, all points $T$ on the radical axis again have the property that we have $TU \cdot TV = TW \cdot TX$, no matter where point $T$ is chosen on the radical axis, where point $U$ is chosen on circle $O$, or where point $W$ is chosen on circle $Q$.

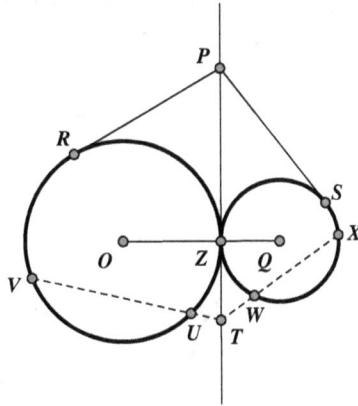

Figure 99a

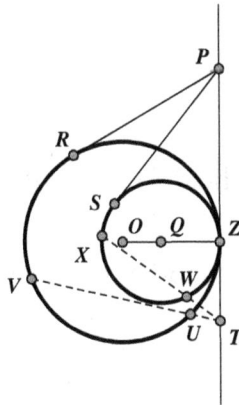

Figure 99b

## Curiosity 100. The Radical Axis of External Non-Intersecting Circles

In Figure 100, we consider two circles with centers $O$ and $Q$ that do not intersect and are external to each other. All points $P$ with the property that the tangents from $P$ to the two circles are of equal length lie on a line, which is the *radical axis* of the two circles. This line lies between the two circles and is perpendicular to line $OQ$. Also, the points $T$ on the radical axis have the property that the equality $TU \cdot TV = TW \cdot TX$ holds for all secants $TUV$ of circle $O$ and all secants $TWX$ of circle $Q$. (This is not illustrated here in order to keep the figure simple, but Figures 98 and 99a, b show this property.)

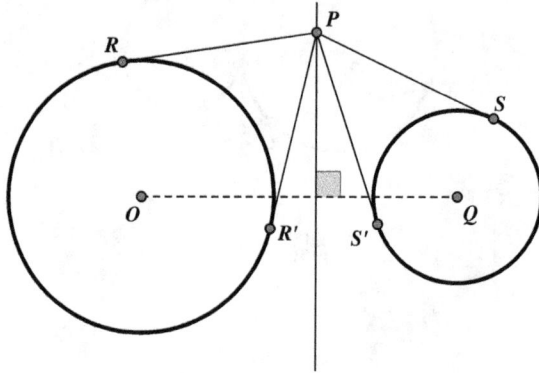

Figure 100

## Curiosity 101. The Radical Axis of Internal Non-Intersecting Circles

In Figure 101, we consider another possible configuration of two circles with centers $O$ and $Q$. Here, the circles do not intersect, and one is internal to the other, with circle $Q$ inside circle $O$. All properties are as in Curiosity 100. Points $P$ with the property that the tangents from $P$ to the two circles are of equal length lie on the *radical axis* of the two circles. As before, the points $T$ on the radical axis also have the property that the equality $TU \cdot TV = TW \cdot TX$ holds for secants $TUV$ of circle $O$ and all secants $TWX$ of circle $Q$. (Again, this is not illustrated here in order to keep the figure simple.)

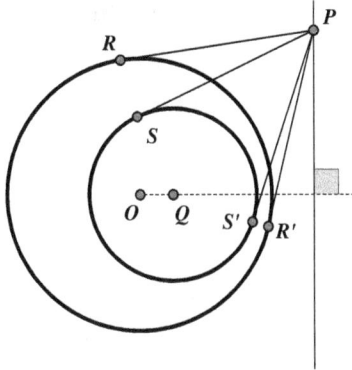

Figure 101

## Curiosity 102. A Coincidence of Intersecting Circles

As introduced in Curiosity 98, the line joining the intersection points of two intersecting circles is called their radical axis. Astonishingly, we can find a connection between the radical axis of two circles created in a triangle and the orthocenter of the triangle. We see this in Figure 102, where points *F* and *G* are on sides *BC* and *CA* of triangle *ABC*, respectively. Lines *AF* and *BG* are diameters of circles with centers *O* and *Q*, respectively, and these circles intersect at points *K* and *L*. Surprisingly, the radical axis *KL* of these circles contains the orthocenter *H* of triangle *ABC*.

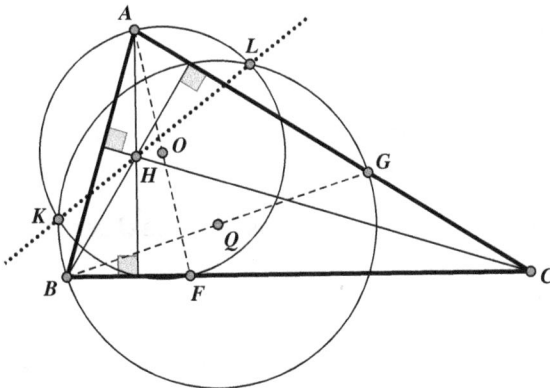

Figure 102

## Curiosity 103. The Radical Center of Three Circles

The radical axes of three given circles whose centers are not collinear are concurrent. We see this illustrated in Figure 103, where the circles with centers $O$, $Q$, and $R$ determine three radical axes that intersect at point $P$. This point $P$ is the center of the circle $c$ that intersects all three circles at right angles, solving a mathematical problem commonly known as *Monge's problem*.

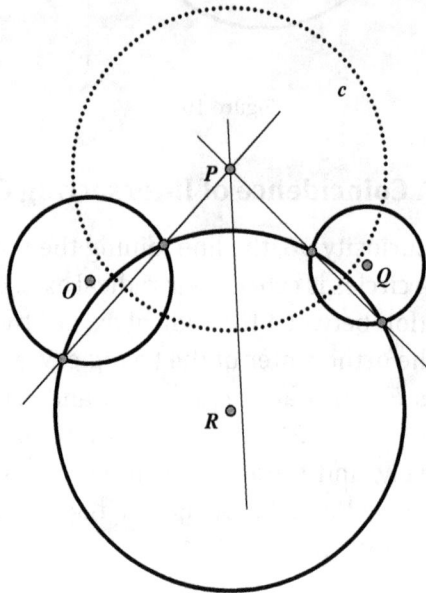

Figure 103

## Curiosity 104. A Surprising Result from a Randomly Selected Point Outside a Circle

Point $P$ on the extension of the radius of circle $O$ is randomly selected external to the circle, as shown in Figure 104. Point $P'$ is then selected along $OP$ so that $OP \cdot OP' = r^2$, where $r$ is the radius of the circle. Point $A$ is randomly selected on circle $O$. Astonishingly, we then find that triangles $OAP$ and $OAP'$ are similar.

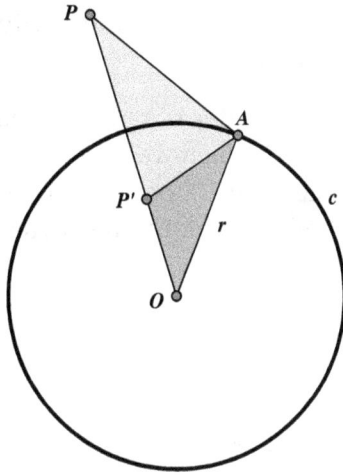

Figure 104

## Curiosity 105. Unexpected Points on a Circle's Tangent

In Figure 105, we are given a circle $c$ with center $O$, radius $r$, and a point $A$ on its circumference. When any point $P$ on the circle with diameter $OA$ is selected, we can determine the point $P'$ on $OP$ extended beyond the circle such that $OP \cdot OP' = r^2$. We then unexpectedly find that the point $P'$ always lies on the tangent $t$ to circle $c$ at point $A$.

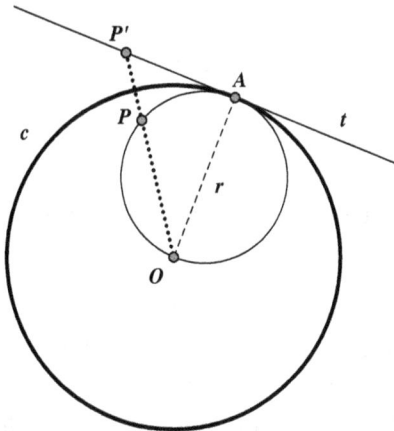

Figure 105

## Curiosity 106. Points on a Secant Generate Points on a Special Circumcircle

In Figure 106, circle $c$ has center $O$, radius $r$, and secant line $s$ intersecting it at points $A$ and $B$. When we choose any point $P$ on the line $s$, we can determine the point $P'$ on the extension of $OP$ such that $OP \cdot OP' = r^2$. We then find that point $P'$ always lies on the circumcircle of triangle $OBA$.

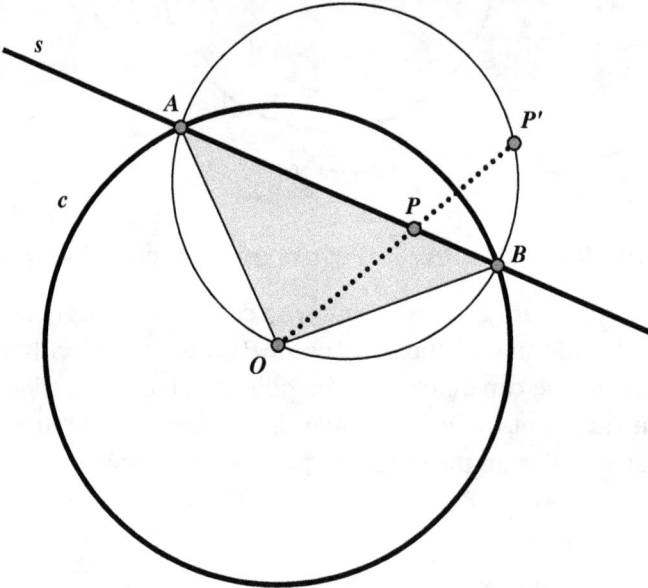

Figure 106

## Curiosity 107. An External Circle Generates an Internal Circle

In Figure 107, we are given a circle $c$ with center $O$ and radius $r$, and another circle $d$ exterior to circle $c$. Incredibly, when random points $P$ of circle $d$ are related to points $P'$ on $OP$ so that the relationship $OP \cdot OP' = r^2$ holds, all the resulting points $P'$ lie on a common circle $d'$, in the interior of circle $c$.

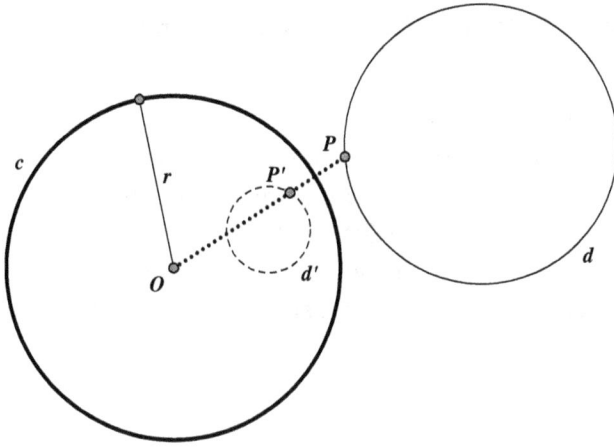

Figure 107

## Curiosity 108. Two Circles Generating Concyclic Points

In Figure 108, we are given a circle $c$ with center $O$ and radius $r$. Inside of circle $c$ is another circle $d$, which contains point $O$ in its interior. When random points $P$ of circle $d$ are related to points $P'$ on $OP$ so that the relationship $OP \cdot OP' = r^2$ holds, all the resulting points $P'$ lie on a common circle $d'$, which contains circle $c$ in its interior.

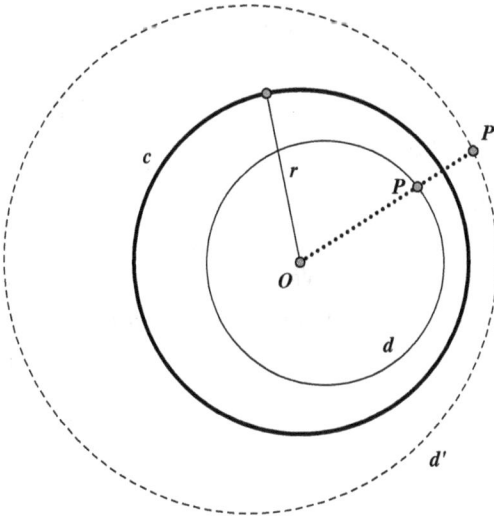

Figure 108

## Curiosity 109. Generating Semicircles on the Sides of a Square

In Figure 109, square *ABCD* is inscribed in circle *c*, which has center *O* and radius *r*. We select a random point *P* on a side of square *ABCD*. Next, we select a point *P'* on the extension of *OP* beyond the circle so that the relationship $OP \cdot OP' = r^2$ holds. For all the points *P* along the sides of square *ABCD*, the points *P'* determine the four semicircles with diameters *AB*, *BC*, *CD*, and *DA* shown in the figure.

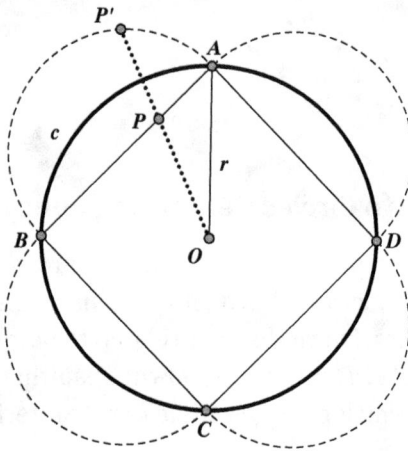

Figure 109

## Curiosity 110. Two Circles Generating an Unexpected Equilateral Triangle

When two circles of equal size contain each other's center, a line drawn from one intersection point of the two circles will always generate an equilateral triangle. We see this in Figure 110, where we have two equal size circles with centers *O* and *Q*. Point *O* lies on circle *Q* and point *Q* lies on circle *O*, and the circles intersect at points *A* and *D*. A line is drawn from the intersection point *D* that cuts the circles *Q* and *O* at points *B* and *C*, respectively. We find that *ABC* is an equilateral triangle.

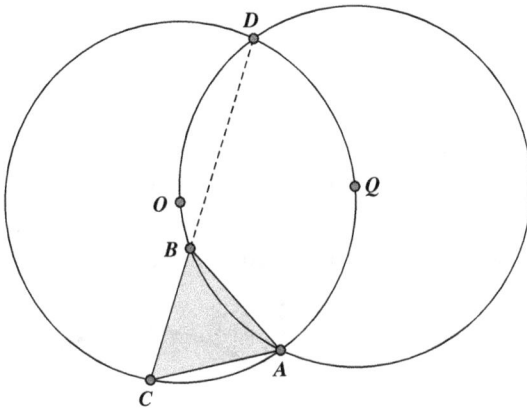

Figure 110

## Curiosity 111. Two Equal Circles Produce an Isosceles Triangle

In Figure 111, two equal circles with centers $O$ and $Q$ intersect at points $A$ and $B$. A line drawn through point $A$ intersects circle $O$ at a point $X$, external to circle $Q$, and intersects circle $Q$ at a point $Y$, external to circle $O$. Amazingly, we find that triangle $BYX$ is always isosceles, no matter how we draw the line through $A$.

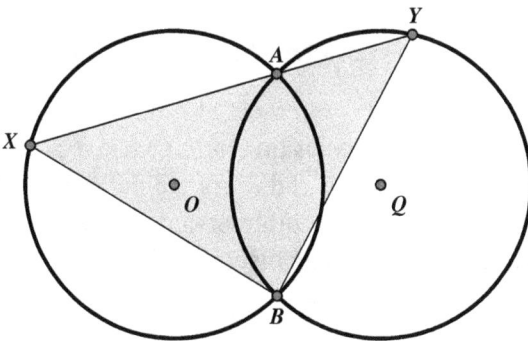

Figure 111

## Curiosity 112. Two Equal Circles Generate Another Isosceles Triangle

In Figure 112, two equal circles with centers $O$ and $Q$ intersect at points $A$ and $B$. A line through point $A$ intersects circle $O$ at a point $X$ (in the interior of circle $Q$), and intersects circle $Q$ at a point $Y$, external to circle $O$. In this configuration, we also find that triangle $BYX$ is isosceles, independent of the choice of the line through $A$.

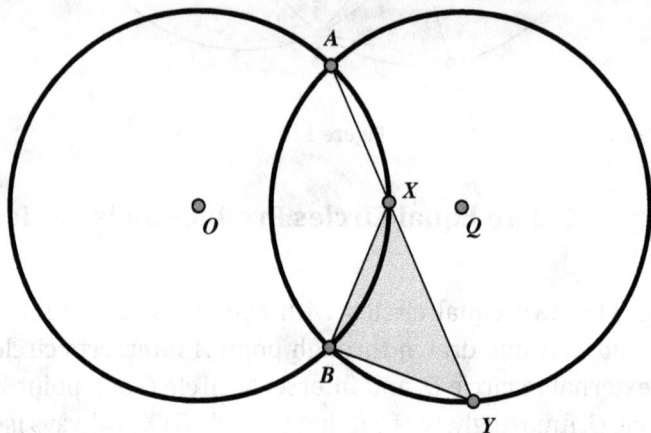

Figure 112

## Curiosity 113. An Unexpected Angle Bisector

Circles with centers $O$ and $Q$ intersect at points $A$ and $B$, as shown in Figure 113. A point $C$ is chosen on circle $O$ and a point $D$ is chosen on circle $Q$ so that points $C$, $B$, and $D$ are collinear. Line $AD$ intersects circle $O$ at point $E$ and line $CE$ intersects circle $Q$ at points $X$ and $Y$. Surprisingly, $AD$ is then the angle bisector of $\angle XAY$.

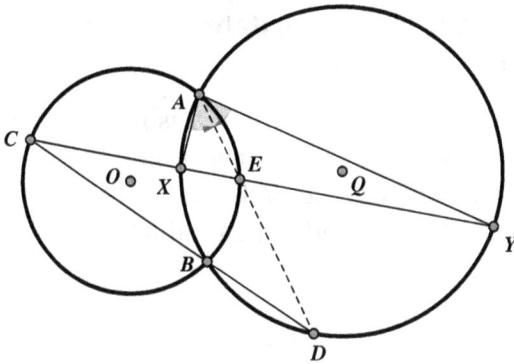

Figure 113

## Curiosity 114. An Amazing Angle Equality in Intersecting Circles

Two circles with centers $O$ and $Q$ intersect at points $X$ and $Y$, as shown in Figure 114. A random line intersects circle $O$ at points $A$ and $B$, and circle $Q$ at points $C$ and $D$. Surprisingly, no matter how the line is chosen, we then have $\angle AXC = \angle DYB$.

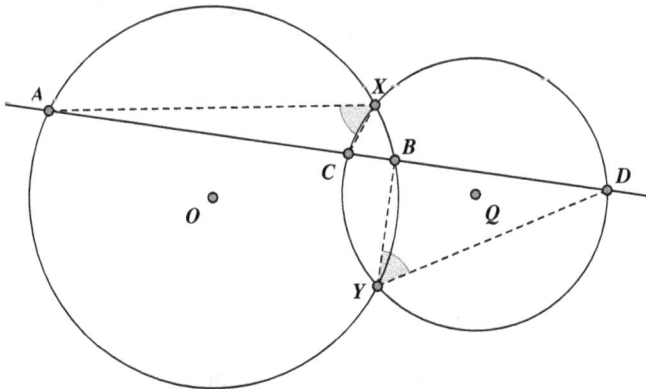

Figure 114

## Curiosity 115. A Cyclic Quadrilateral Generates Two Circles with Perpendicular Radii

In Figure 115, cyclic quadrilateral *ABCD* is inscribed in a circle with center *O* and its extended sides meet at points *E* and *F* as shown. Segment *EF* is a diameter of the circle with center *M*, which intersects circle *O* at point *N*. Unexpectedly, the radii *NO* and *NM* of the two circles are perpendicular.

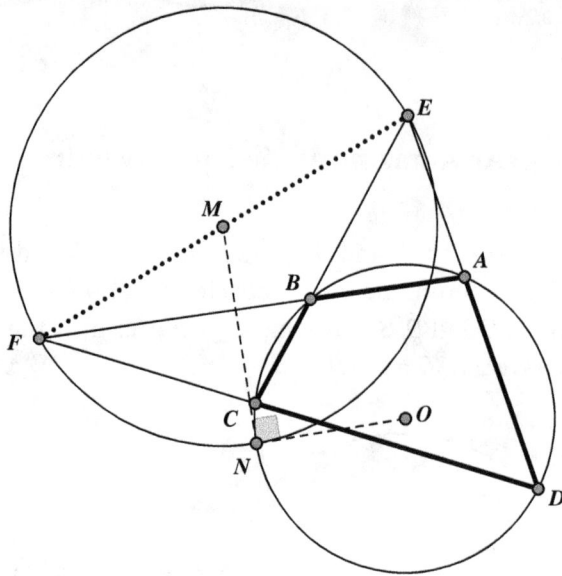

Figure 115

## Curiosity 116. Circles That Generate an Equality of Seemingly Unrelated Segments

In Figure 116, circle *O* has *CD* as its diameter. Point *D* is the midpoint of arc *AB*. Chord *CE* extended intersects chord *AB* extended at point *F*. A circle with center *C* contains the midpoint *H* of *EF* and intersects *BD* at point *G*. Quite unexpectedly, we find that *BG* = *HE* = *HF*.

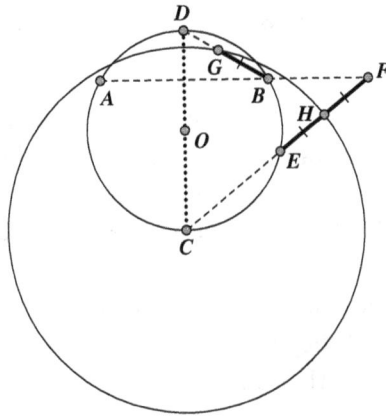

Figure 116

## Curiosity 117. Related Equilateral Triangles Inscribed in Concentric Circles

A curious relationship evolves when two equilateral triangles are each inscribed in concentric circles. In Figure 117, triangle *ABC* is inscribed in the larger circle *O*, and triangle *A′B′C′* is inscribed in the smaller circle *O*. Point *P* lies on the larger circle and point *P′* lies on the smaller circle. The sum of the squares of *PA′*, *PB′*, and *PC′* is equal to the sum of the squares of *P′A*, *P′B*, and *P′C*. Symbolically, we have $(PA')^2 + (PB')^2 + (PC')^2 = (P'A)^2 + (P'B)^2 + (P'C)^2$.

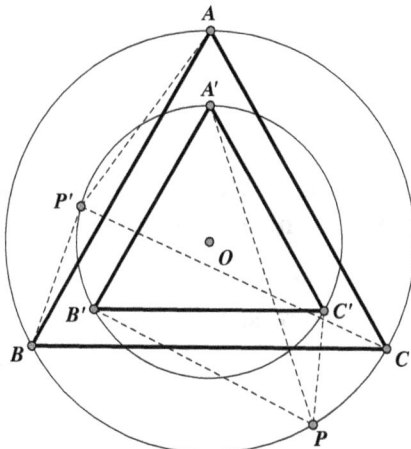

Figure 117

## Curiosity 118. Intersecting Circles Surprisingly Generate Equal Lines

We are faced here with a curious situation involving two intersecting circles, one of which contains the center of the other. Such circles can be shown to generate two intersecting lines that always turn out to be equal in length. This can be seen in Figure 118, where the circle with center *A* contains the center of circle *B*. The circles intersect at points *C* and *D*. When we draw two chords, *CE* and *DF*, in the circle with center *B* such that they intersect at a common point *P* on circle *A*, the unexpected result is that *CE* = *DF*.

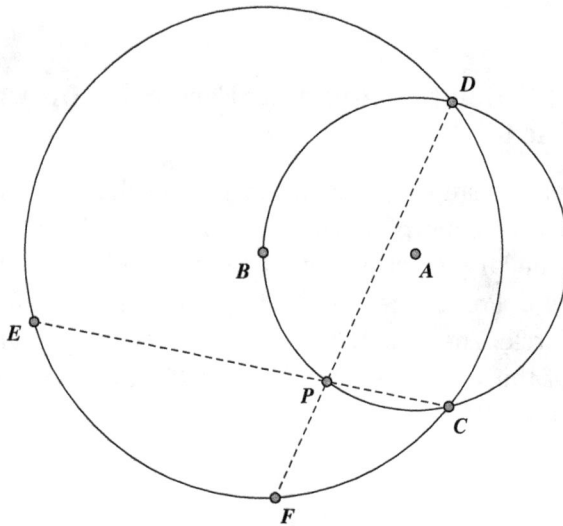

Figure 118

## Curiosity 119. Two Intersecting Circles Produce an Unexpected Line Segment Relationship

In Figure 119, a circle with center *Q* contains the center *O* of a second circle. Point *A* is a point of intersection of the two circles, and diameter *AQB* of circle *Q* intersects circle *O* at point *C*. Strangely, we find that $2OC^2 = AB \cdot AC$.

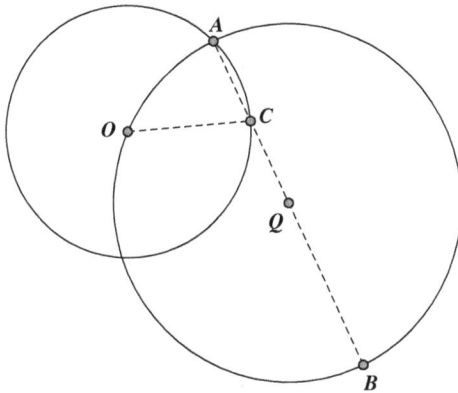

Figure 119

## Curiosity 120. A Quadrilateral with Two Intersecting Circles Generates Yet Another Circle

In Figure 120, quadrilateral *ABCD* is partitioned into two triangles *ABC* and *ACD*, each of which is circumscribed by a circle. The circumscribed circle about triangle *ABC* intersects *DC* and *DA* at points *E* and *F*, respectively, and the circumscribed circle about triangle *ACD* intersects *BA* and *BC* at points *P* and *Q*, respectively. Lines *BF* and *BE* intersect *PQ* at points *R* and *S*, respectively. Quite astonishingly, the points *E*, *F*, *R*, and *S* also turn out to be concyclic.

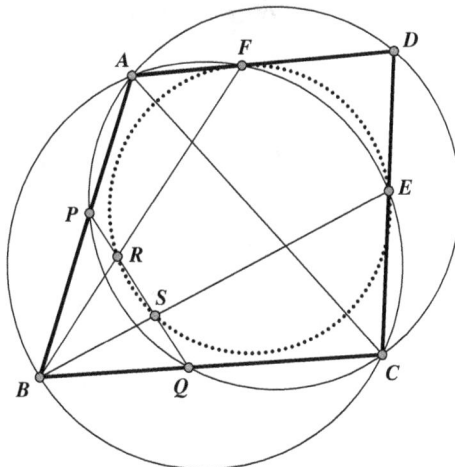

Figure 120

## Curiosity 121. The Broken Chord

In Figure 121, we consider triangle $ABC$, which is inscribed in circle $O$. Point $M$ is the midpoint of arc $BAC$, and the perpendicular from point $M$ to $AB$ intersects it at point $E$. Curiously, it turns out that $BE = AE + AC$.

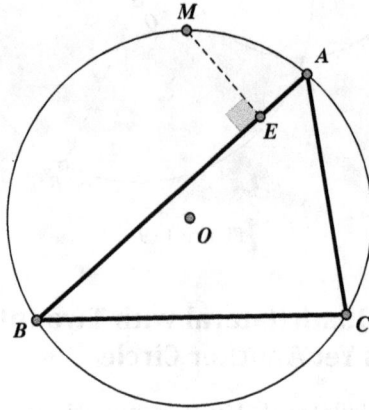

Figure 121

## Curiosity 122. Tangents Yield Equal Segments

Circles with centers $O$ and $Q$ intersect at points $A$ and $B$, as shown in Figure 122. Point $C$ is the intersection of circle $O$ and the tangent to circle $Q$ at point $A$, and point $D$ is the intersection of circle $Q$ and the tangent of circle $O$ at point $A$. Point $P$ is the intersection of the extension of $CB$ with circle $Q$. Surprisingly, we then have $AD = AP$.

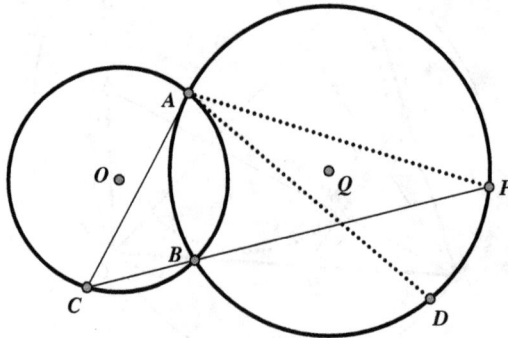

Figure 122

## Curiosity 123. The Fourth Corner of a Rectangle Lies on a Special Circle

In Figure 123, two circles with centers $O$ and $Q$ intersect at points $A$ and $B$. A line through $B$ and parallel to $OQ$ intersects circles $O$ and $Q$ at points $X$ and $Y$, respectively. When we draw any rectangle $ADEF$ with vertex $D$ on circle $O$ and vertex $F$ on circle $Q$, we find that point $E$ must lie on the circle with diameter $XY$.

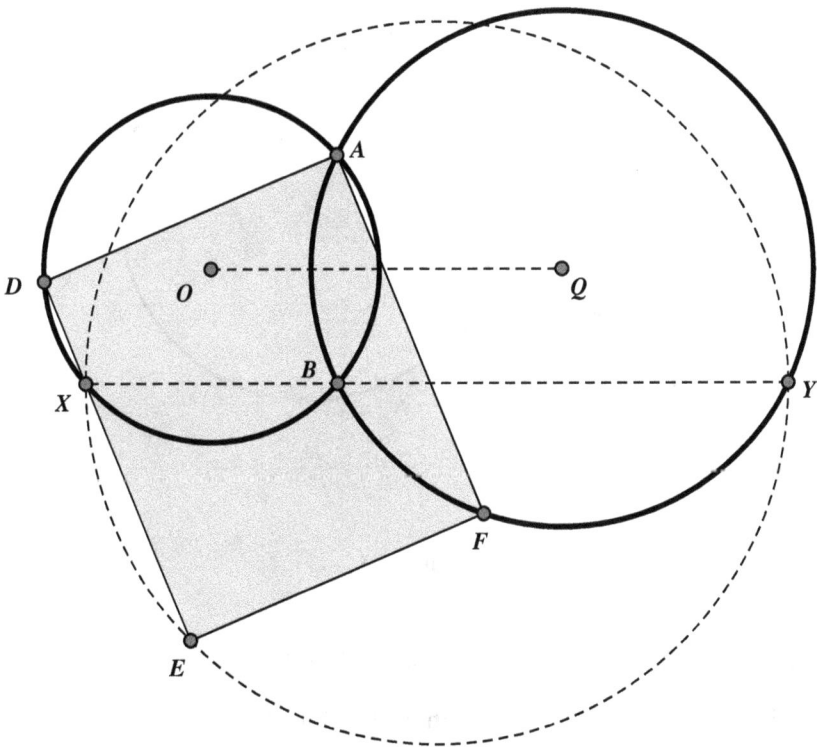

Figure 123

## Curiosity 124. Creating Equal Chords in Two Circles

In Figure 124, we are given two nonintersecting circles with centers $O$ and $Q$. Point $M$ is the midpoint of $OQ$ and thus the center of the circle with diameter $OQ$. This circle intersects circle $O$ at point $A$, and intersects circle $Q$ at point $B$, whereby points $A$ and $B$ lie on the same side of line $OQ$. The extension of line $AB$ intersects circle $O$ a second time at point $C$ and intersects circle $Q$ a second time at point $D$. Unexpectedly, we find that $AC = BD$.

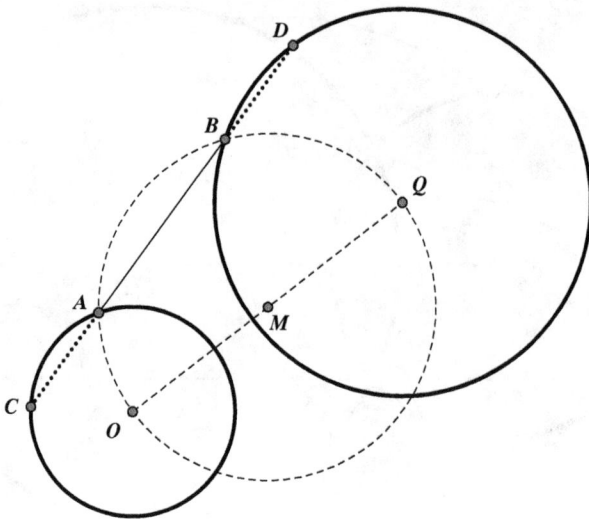

Figure 124

## Curiosity 125. An Unusual Angle Bisector Construction

In this Curiosity, we encounter an unusual method to bisect the angle in a triangle. In Figure 125, we are given triangle $ABC$ with points $D$ and $E$ on sides $BC$ and $AC$, respectively, so that $BD = AE$. The circumcircles of $ADC$ and $BCE$ intersect at an external point $P$. Surprisingly, $CP$ is the bisector of angle $\angle ACB$.

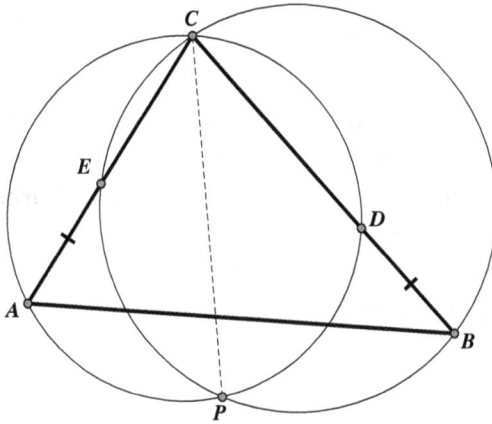

Figure 125

## Curiosity 126. A Square Derived from Two Circles with a Surprising Area

We are given two points $A$ and $B$, as shown in Figure 126. A circle with center $O$ and radius $r$ is tangent to line $AB$ at point $A$, and a circle with center $Q$ and radius $s$ is tangent to line $AB$ at point $B$. The other intersection point of these two circles is point $C$. Point $M$ is the circumcenter of triangle $ABC$. When we draw a square with side $CM$, we find that its area is equal to the product $r \cdot s$ of the radii of the circles $O$ and $Q$.

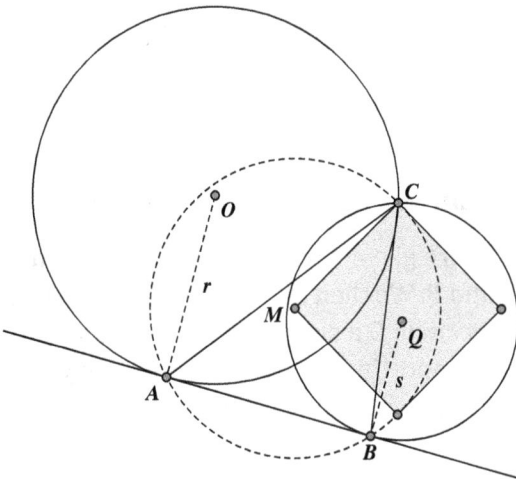

Figure 126

## Curiosity 127. An Unexpected Circumcircle Center

A circle with center $O$ contains four points $A$, $B$, $C$, and $D$ as shown in Figure 127, where $AC$ is a diameter of the circle. The circle with center $B$ and radius $AD$ intersects the circle with center $D$ and radius $AB$ at a point $P$. Circle $B$ intersects circle $O$ at point $Q$, and circle $D$ intersects circle $O$ at point $R$. The points $P$, $Q$, and $R$ all lie on the side of line $BC$ opposite that of point $A$. We then find that point $C$ is the center of a circle containing points $P$, $Q$, and $R$, and we have $CP = CQ = CR$.

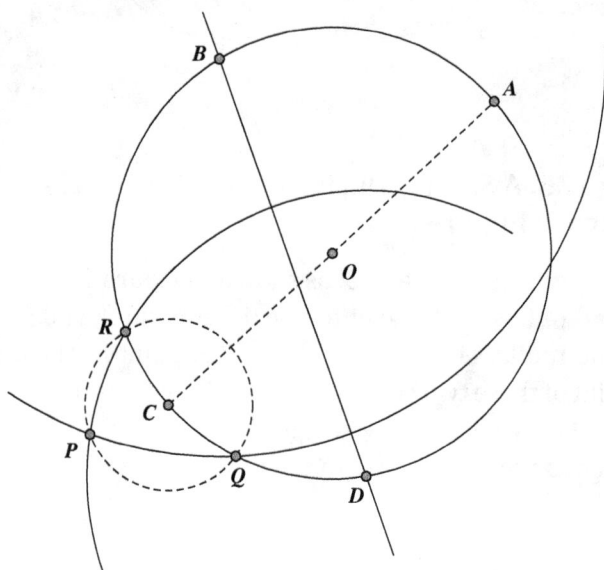

Figure 127

## Curiosity 128. Chords of a Constant Length

In Figure 128, we are given circles with centers $O$ and $Q$, which intersect at points $A$ and $B$. We then choose a point $P$ at random on circle $Q$. Line $PA$ intersects circle $O$ a second time at point $X$, and line $PB$ intersects circle $O$ a second time at point $Y$. Amazingly, we find that the length of line $XY$ has a constant value, independent of the choice of point $P$.

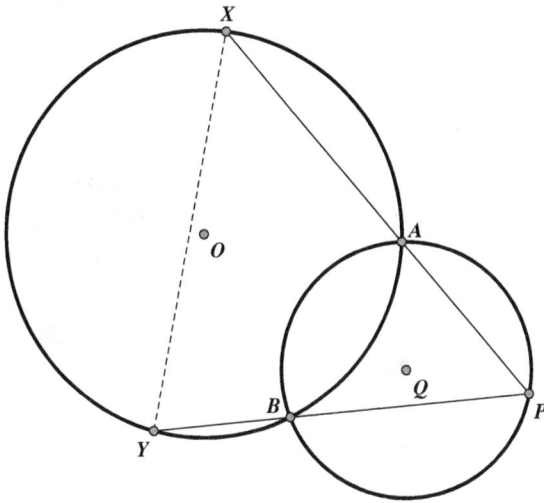

Figure 128

## Curiosity 129. An Unexpected Equality of Two Special Chords

In Figure 129, we are given circles with centers $O$ and $Q$ with point $Q$ on circle $O$. These circles intersect at points $A$ and $B$. The tangent of circle $O$ at point $A$ intersects circle $Q$ at point $C$. We find that the common chord $AB$ is the same length as the tangent segment $AC$; in other words, $AB = AC$.

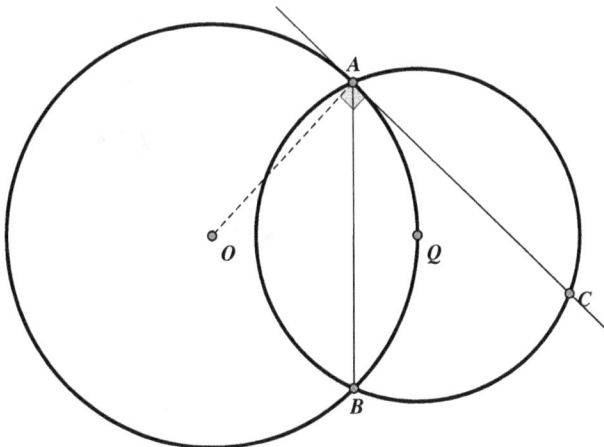

Figure 129

## Curiosity 130. A Surprising Common Point of Three Circles

In Figure 130, we are given three mutually external circles with centers $O$, $P$, and $Q$. The common internal tangents of circles $P$ and $Q$ intersect at point $A$, the common internal tangents of circles $Q$ and $O$ intersect at point $B$, and the common internal tangents of circles $O$ and $P$ intersect at point $C$. We find that lines $OA$, $PB$, and $QC$ are concurrent at point $E$.

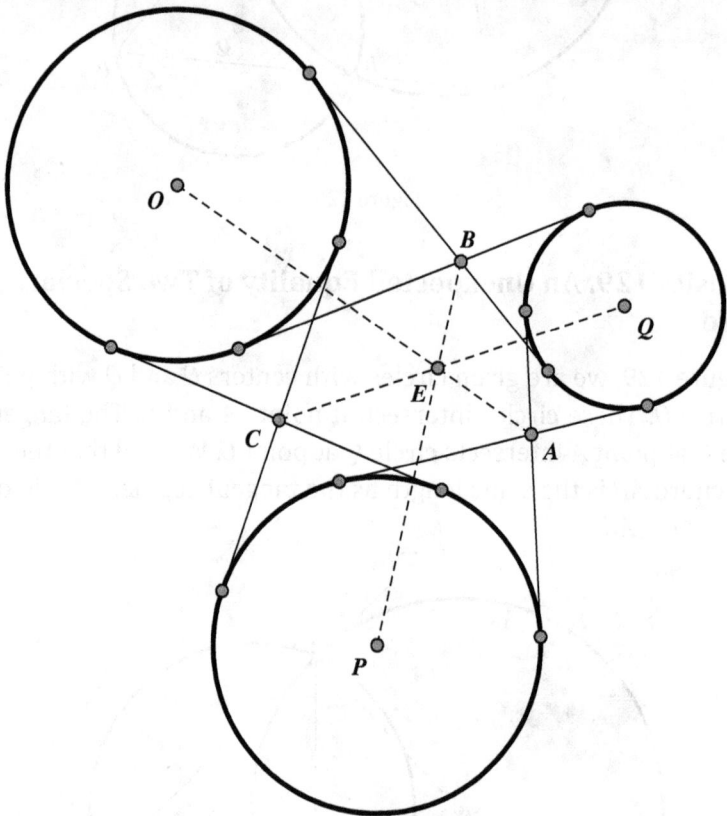

Figure 130

## Curiosity 131. Perpendicular Circles Create a Diameter

In Figure 131, two circles with centers $O$ and $Q$ intersect at right angles at points $A$ and $B$. Point $P$ is chosen anywhere on circle $Q$ in the interior of circle $O$. Point $X$ is the intersection of $AP$ with circle $O$ and point $Y$ is the intersection of $BP$ with circle $O$. We then find that $XY$ is the diameter of circle $O$.

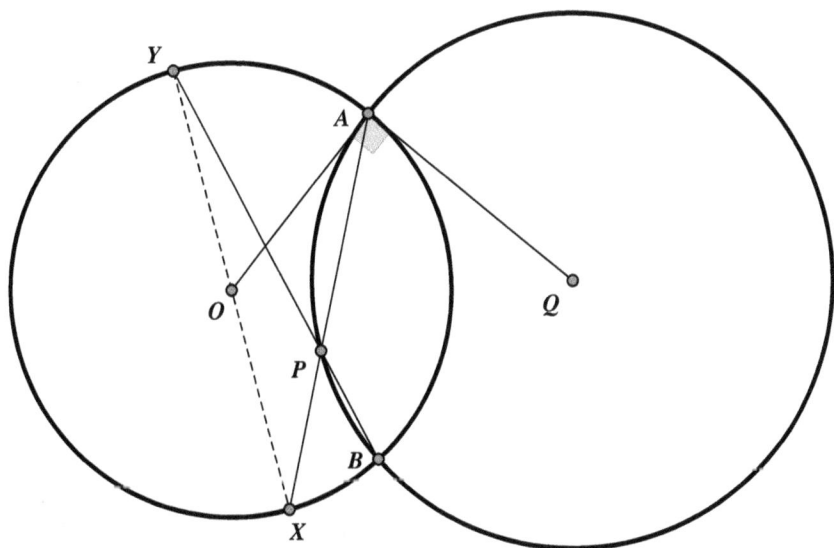

Figure 131

# Proofs of the Circle Curiosities

### Curiosity 1. A Triangle Defines a Circle

It is common knowledge that the three vertices of a triangle uniquely define the center of the triangle's circumcircle, but it is well worth recalling why this is the case. As we see in Figure 1-P, all points that are the same distance from vertices $A$ and $B$ lie on the perpendicular bisector $f$ of segment $AB$, containing the midpoint $F$ of $AB$. Similarly, all points that are the same distance from vertices $B$ and $C$ lie on the perpendicular bisector $d$ of segment $BC$, containing the midpoint $D$ of $BC$. Lines $f$ and $d$ intersect in a unique point $O$, which has the property $OA = OB$ (since it is a point on $f$) and $OB = OC$ (since it is a point on $d$). Thus, $OA = OB = OC$.

We note that since $OC = OA$, point $O$ also lies on the perpendicular bisector $e$ of segment $CA$. Point $O$ is therefore a common point of the perpendicular bisectors of all three sides of triangle $ABC$, which is then the center of the circle circumscribed about triangle $ABC$.

It is worthwhile to review some of the elementary principles from school geometry, since the *why* is often just as fascinating as the *what*!

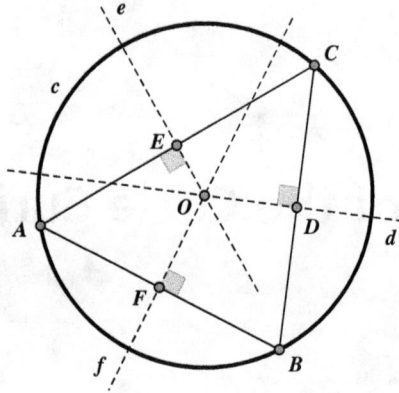

Figure 1-P

## Curiosity 2. A Circle Tangent is Perpendicular to the Radius in the Point of Tangency

Why, exactly, does the line perpendicular to the radius at point $A$ of a circle only have that one point in common with the circle? In Figure 2-P, we assume the existence of a second point $B \neq A$ that is on both the circle and the line through $A$ perpendicular to $OA$. If such a point $B$ exists, $OBA$ is a right triangle with hypotenuse $OB$. By the Pythagorean theorem, we then have $OB^2 = OA^2 + AB^2$. Since $AB > 0$, we then have $OB > OA = r$, which contradicts the assumption that point $B$ lies on the circle with center $O$ and radius $r$. We see that such a second point cannot exist, and the line perpendicular to $OA$ is therefore the tangent of the circle in $A$.

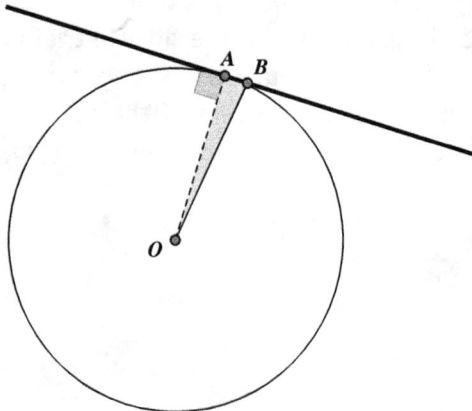

Figure 2-P

## Curiosity 3. Circle Tangents from a Common Point are of Equal Length

This Curiosity is just a consequence of the symmetry of the configuration with respect to the line $OP$. If we wish to give a more concrete proof for the equality of the tangents, we consider the triangles $PAO$ and $POB$, as shown in Figure 3-P. Noting that $\angle OAP = \angle PBO = 90°$ follows from Curiosity 2 and $OA = OB = r$, we see that these triangles are both right triangles with the common hypotenuse $OP$ and a side of length $r$. They are therefore congruent, and we have $PA = PB$.

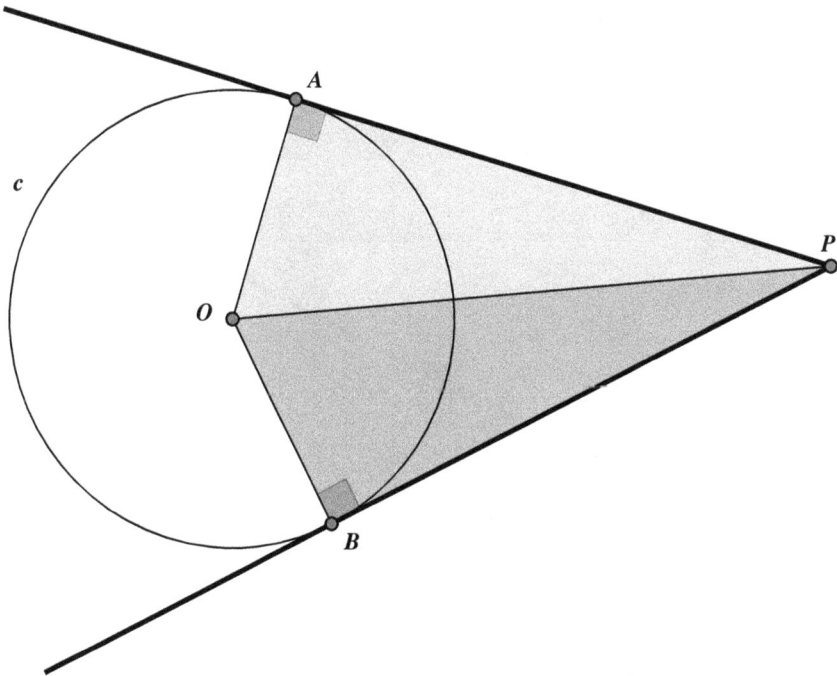

Figure 3-P

## Curiosity 4. The Angle Between Circle Tangents from a Common Point is Supplementary to its Intercepted Arc

In Figure 4-P, we note that we can write $\angle APB = \angle P = \frac{1}{2}\left(\text{large } \overset{\frown}{AB} - \text{small } \overset{\frown}{AB}\right) = \frac{1}{2}\left((360° - \angle BOA) - \angle BOA\right) = 180° - \angle BOA$. This shows us the equivalence of the two notations. We know from Curiosity 3 that $\angle OAP = \angle PBO = 90°$, and considering quadrilateral $OBPA$ then gives us $\angle APB = 360° - \angle OAP - \angle BOA - \angle PBO = 360° - 90° - \angle BOA - 90° = 180° - \angle BOA$.

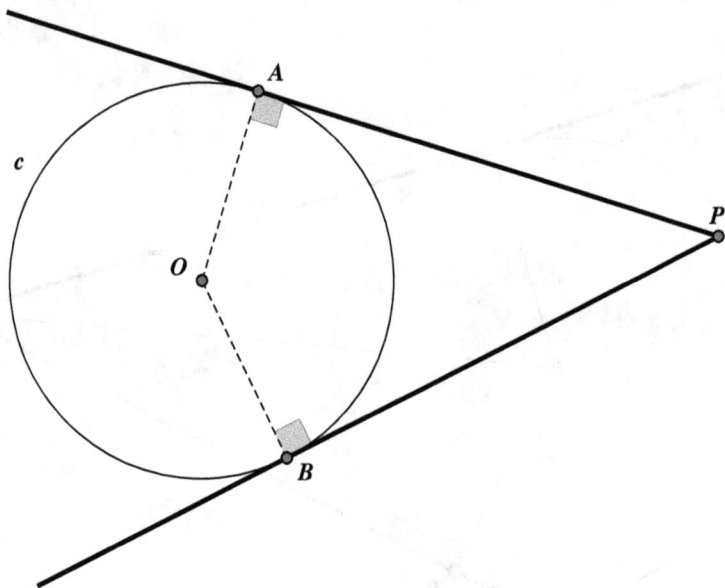

Figure 4-P

## Curiosity 5. Angle Between a Circle Tangent and a Chord

This is an immediate consequence of the fact that triangle $OBA$ is isosceles. In Figure 5-P, we see that $OA$ and $OB$ are both radii of circle $c$. This means that we have $\angle OAB = \angle ABO$, and since we know that $\angle PBO = 90°$, we obtain $\angle PBA = 90° - \angle ABO = \frac{1}{2}(180° - 2\angle ABO) = \frac{1}{2}\angle BOA = \frac{1}{2}\overset{\frown}{AB}$.

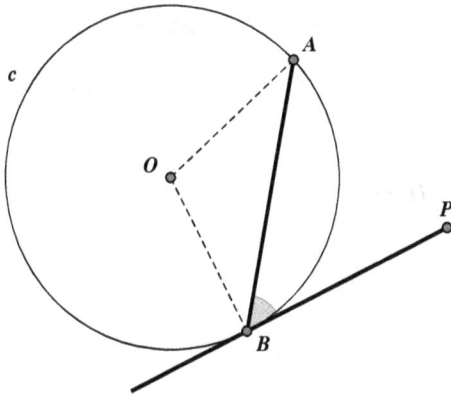

Figure 5-P

## Curiosity 6. The Angle Formed by Two Chords of a Circle

In Figure 6-P, we add the chords $AC$, $CB$, and $BD$. Since both angles $\angle BAC$ and $\angle BDC$ are subtended on chord $BC$, they are equal. Since vertical angles $\angle CPA$ and $\angle DPB$ are also equal, we know that triangles $PCA$ and $PDB$ are similar. From this, we obtain $\frac{AP}{CP} = \frac{DP}{BP}$, which is equivalent to $AP \cdot BP = CP \cdot DP$.

Now, considering triangle $PBC$, we see that exterior angle $\angle CPA = \angle CBP + \angle PCB$. Thus, since $\angle CBP = \angle CBA = \frac{1}{2}\angle COA$ and $\angle PCB = \angle DCB = \frac{1}{2}\angle DOB$, we can obtain $\angle CPA = \frac{1}{2}\angle COA + \frac{1}{2}\angle DOB = \frac{1}{2}(\angle COA + \angle DOB)$ or $\angle CPA = \frac{1}{2}(\overset{\frown}{AC} + \overset{\frown}{BD})$.

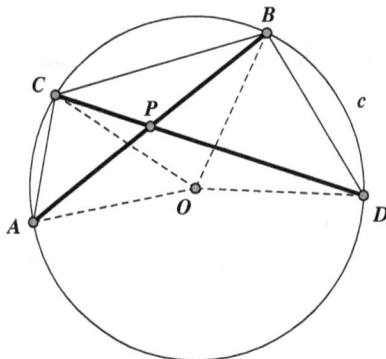

Figure 6-P

## Curiosity 7. Surprising Supplementary Angles

In Figure 7-P, we first draw chords $AD$, $DB$, and $BC$. On chord $AD$, we have $\angle AOD = 2\angle ABD$, since both angles are measured by arc $AD$. Similarly, on chord $BC$, we have $\angle BOC = 2\angle BDC$. Also, in right triangle $PDB$, we have $\angle PBD = 90° - \angle BDP$. In summary, we therefore obtain

$$
\begin{aligned}
\angle AOD &= 2 \cdot \angle ABD \\
&= 2 \cdot \angle PBD \\
&= 2 \cdot (90° - \angle BDP) \\
&= 180° - 2 \cdot \angle BDC \\
&= 180° - 2 \cdot \frac{1}{2} \cdot \angle BOC \\
&= 180° - \angle BOC.
\end{aligned}
$$

We see that angles $\angle AOD$ and $\angle BOC$ are supplementary, as we set out to prove.

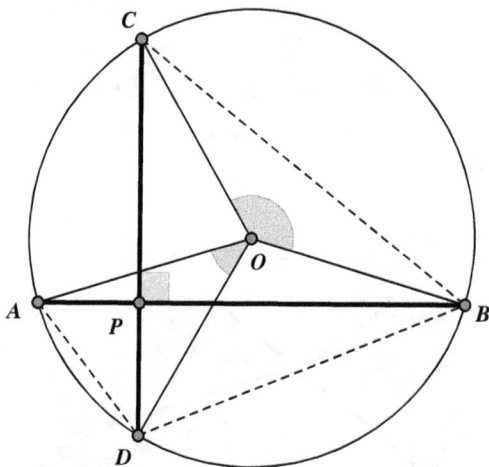

Figure 7-P

## Curiosity 8. The Angle Formed by a Tangent and a Secant

In Figure 8-P, we add the chords $AB$ and $AC$. Since the subtended angle $\angle BCA$ on chord $AB$ is equal to $\angle BAP$, we see that triangles $PCA$ and $PDB$ are similar, and thus, $\frac{PB}{PA} = \frac{PA}{PC}$.

Now, considering triangle $PAB$, we see that $\angle APC = \angle ABC - \angle BAP$. Since $\angle ABC = \frac{1}{2}\angle AOC$ and $\angle BAP = \angle BCA = \frac{1}{2}\angle BOA$, we obtain $\angle APC = \frac{1}{2}\angle AOC - \frac{1}{2}\angle BOA = \frac{1}{2}(\angle AOC - \angle BOA)$ or $\angle APC = \frac{1}{2}\left(\overarc{AC} - \overarc{AB}\right)$.

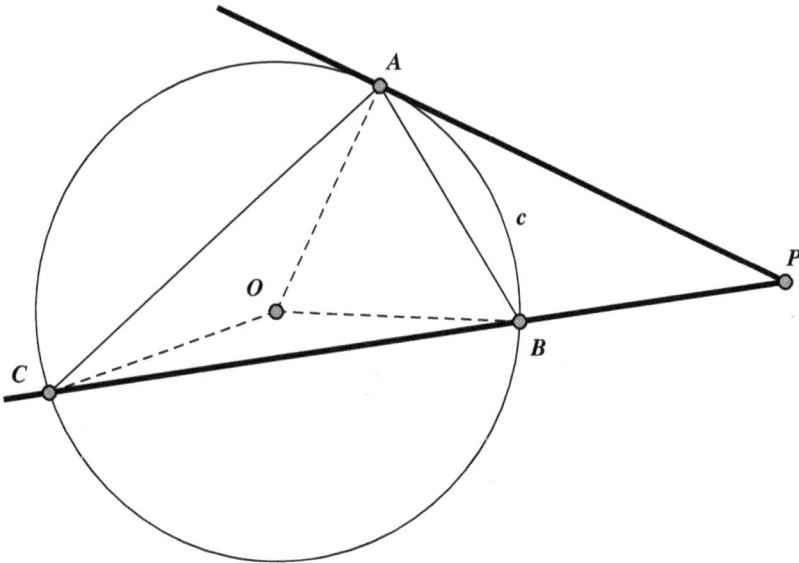

Figure 8-P

## Curiosity 9. The Angle Formed by Two Secants

In Figure 9-P, we add the chords $AD$ and $BC$. Since both angles $\angle BCP = \angle BCD$ and $\angle BAD = \angle PAD$ are subtended on arc $BD$, they are equal. Therefore, triangles $PDA$ and $PCB$ are similar. From this, we obtain $\frac{AP}{CP} = \frac{DP}{BP}$, which is equivalent to $AP \cdot BP = CP \cdot DP$.

Now, considering triangle $PDA$, we see that $\angle CPA = \angle DPA = \angle CDA - \angle PAD = \angle CDA - \angle BAD$. Since $\angle CDA = \frac{1}{2} \angle COA$ and $\angle BAD = \frac{1}{2} \angle BOD$, we obtain $\angle CPA = \frac{1}{2} \angle COA - \frac{1}{2} \angle BOD = \frac{1}{2} (\angle COA - \angle BOD)$ or $\angle CPA = \frac{1}{2} \left( \widehat{AC} - \widehat{DB} \right)$.

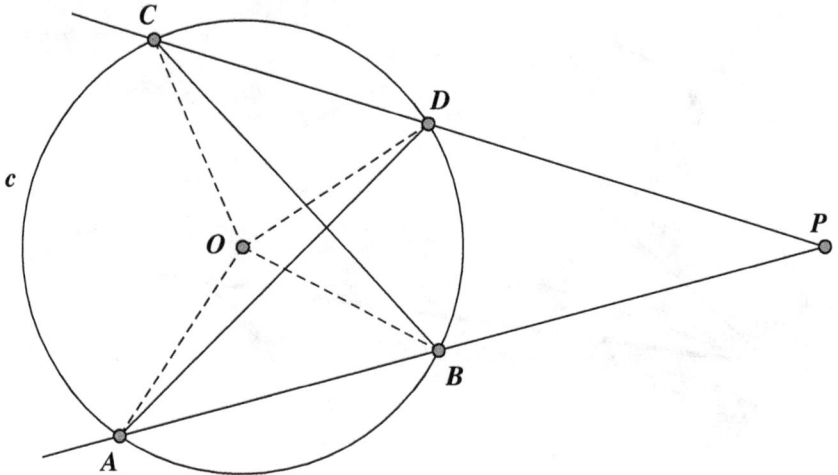

Figure 9-P

## Curiosity 10. Trisecting an Angle Using a Circle

Considering Figure 10-P, we see that triangle $CDB$ is isosceles, with $BC = DC$ and $\angle CBD = \angle BDC$. In triangle $CDB$, the exterior angle gives us $\angle ECD = \angle CBD + \angle BDC = 2\angle CBD$. Since radii $DC = DE$, triangle $DCE$ is also isosceles, and we have $\angle DEC = \angle ECD = 2\angle CBD$. In triangle $DBE$, we then obtain $\angle EDA = \angle DEB + \angle EBD = 2\angle CBD + \angle CBD = 3\angle CBD$ or $\angle CBD = \frac{1}{3} \cdot \angle EDA$, and so, we have trisected angle $\angle EDA$.

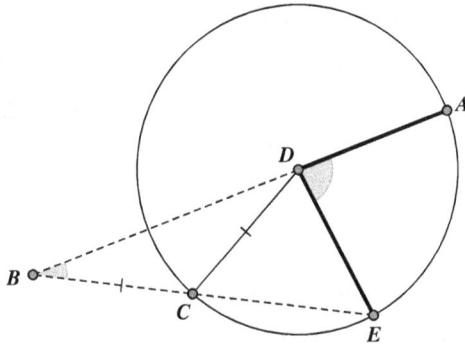

Figure 10-P

## Curiosity 11. A Rather Surprising Equality

In Figure 11-P, we extend tangent $KB$ to meet $LA$ at point $S$, and extend $LA$ its own length beyond $A$ to point $R$. We have tangents $SB = SA$, and since $AL = BK$, we therefore have $SR = SA + AR = SB + AL = SB + BK = SK$. We then have isosceles triangles $SKR$ and $SBA$. Since $\frac{SB}{SK} = \frac{SA}{SR}$, we have $RK||AB$. Considering triangle $KRL$, we have $RK||AP$. Since $A$ is the midpoint of $RL$, the point $P$ must be the midpoint of $KL$.

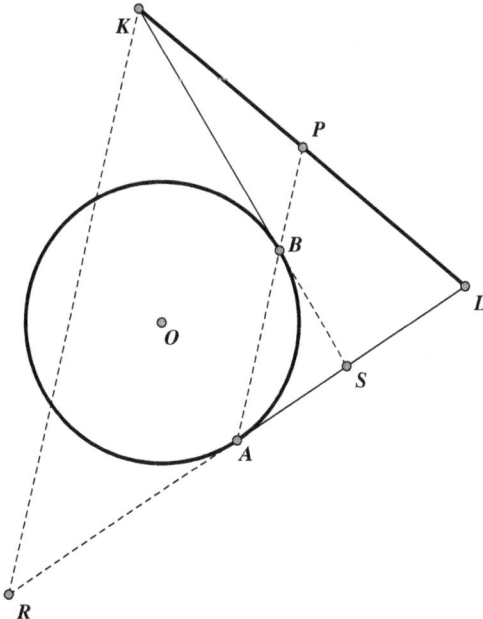

Figure 11-P

## Curiosity 12. A Strange Appearance of Parallel Lines

In Figure 12-P, point $Y$ is the intersection of line $l = BCT$ with $DA$, and we add lines $BA$, $CA$, and $CD$. Since right triangles $\triangle AXT$ and $\triangle ARX$ share a side in $AX$, and we have $XR = XT$, they are congruent, and thus, $\angle XRA = \angle ATX$. We now consider some useful angle equalities. We note that $DB = DC$ implies $\angle BAD = \angle DAC$; therefore, $\angle BAD = \angle BCD$, as both of these inscribed angles are measured by arc $BD$. Furthermore, since $t$ is tangent to circle $c$ at an endpoint of chord $CA$, we have $\angle CAT = \angle CDA$, since both angles are measured by arc $CA$. We then obtain

$$\angle YAT = \angle DAT = \angle DAC + \angle CAT = \angle BAD + \angle CDA =$$
$$\angle BCD + \angle CDA = \angle YCD + \angle CDY = \angle TYA.$$

Right triangles $\triangle AXT$ and $\triangle TXY$ are, thus, congruent, since they share a side in $TX$, and we have $\angle YAT = \angle TYA$. This then gives us $\angle ATX = \angle XTY$. Summing up, we have shown $\angle TRA = \angle XRA = \angle ATX = \angle XTY = \angle RTY$, and since $\angle TRA = \angle RTY$, we see that $AR||TY$ or $AR||l$.

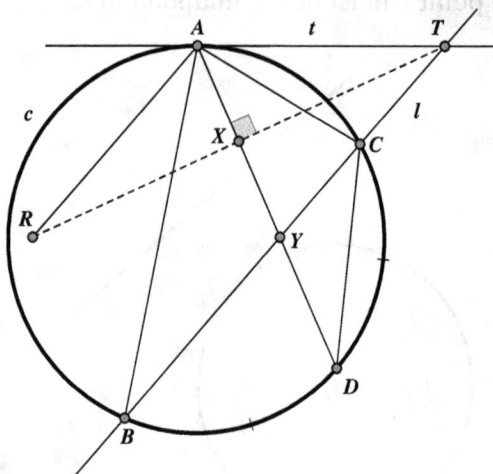

Figure 12-P

## Curiosity 13. Unexpected Complementary Angles in a Circle

When we draw $AC$, as shown in Figure 13-P, we create a right triangle $ABC$, since angle $ACB$ is inscribed in a semicircle. Next, we note that $\angle ADE = \angle ACE$, since both angles are measured by one-half arc $AE$. Since $\angle ACE + \angle BCE = 90°$, we have $\angle ADE + \angle BCE = \angle ACE + \angle BCE = 90°$, which is what we set out to prove.

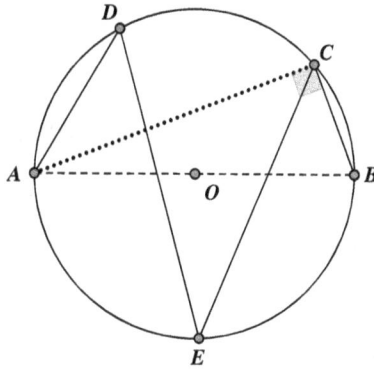

Figure 13-P

## Curiosity 14. Surprising Squares Equality in a Circle

In Figure 14-P, if point $O$ is the center of the given circle, and we are given $\angle ABC = \angle DCB = \angle CDE = 45°$, we have $\angle AOC = \angle DOB = \angle COE = 90°$. This implies that $\angle BOC = \angle AOC + \angle BOA = 90° + \angle BOA$, $\angle COD = 360° - \angle AOC - \angle BOA - \angle DOB = 360° - 90° - \angle BOA - 90° = 180° - \angle BOA$, and $\angle EOD = 180° - \angle BOA - \angle DOB = 180° - \angle BOA - 90° = 90° - \angle BOA$. Since the radii in circle $O$ are all equal, we have $OA = OB = OC = OD = OE = r$. We can now apply the law of cosines (see Toolbox) in various triangles to obtain the following:

For $\triangle OBA$:

$$AB^2 = OA^2 + OB^2 - 2 \cdot OA \cdot OB \cdot \cos\angle BOA = 2r^2 - 2r^2 \cdot \cos\angle BOA.$$

For $\triangle OCD$:

$$CD^2 = OC^2 + OD^2 - 2 \cdot OC \cdot OD \cdot \cos\angle COD =$$
$$2r^2 - 2r^2 \cdot \cos(180° - \angle BOA) = 2r^2 + 2r^2 \cdot \cos\angle BOA.$$

For $\triangle OBC$:

$$BC^2 = OB^2 + OC^2 - 2 \cdot OB \cdot OC \cdot \cos\angle BOC =$$
$$2r^2 - 2r^2 \cdot \cos(90° + \angle BOA) = 2r^2 + 2r^2 \cdot \sin\angle BOA.$$

For $\triangle ODE$:

$$DE^2 = OD^2 + OE^2 - 2 \cdot OD \cdot OE \cdot \cos\angle DOE =$$
$$2r^2 - 2r^2 \cdot \cos(90° - \angle BOA) = 2r^2 + 2r^2 \cdot \sin\angle BOA.$$

Adding the first two equations and then the second two equations yields $AB^2 + CD^2 = 4r^2 = BC^2 + DE^2$, and thus, $AB^2 + CD^2 = BC^2 + DE^2$.

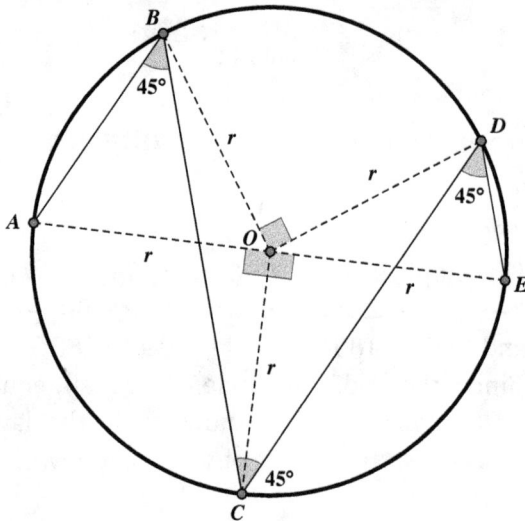

Figure 14-P

## Curiosity 15. A Surprise of a Circle Presenting a Familiar Triangle

In Figure 15-P, we first add lines $OD$ and $OE$. We know that $DE = \frac{1}{2} \cdot AB = AO$ and $OD = OE = OA$, as they are all radii of circle $O$. Consequently, triangle $DOE$ is equilateral with $\angle ODE = 60°$. We are given that $DE$ is parallel to $AB$, and also to $AO$, which gives us $\angle DOA = \angle ODA = 60°$. This establishes triangle $AOD$ as equilateral, as we have $OA = OD$ and $\angle DOA = 60°$. This means that we also have $\angle OAD = \angle BAC = 60°$. Right triangle $ABC$ is therefore the familiar $30° - 60° - 90°$ triangle, where the hypotenuse $AC$ equals twice the length of the shorter side $AB$.

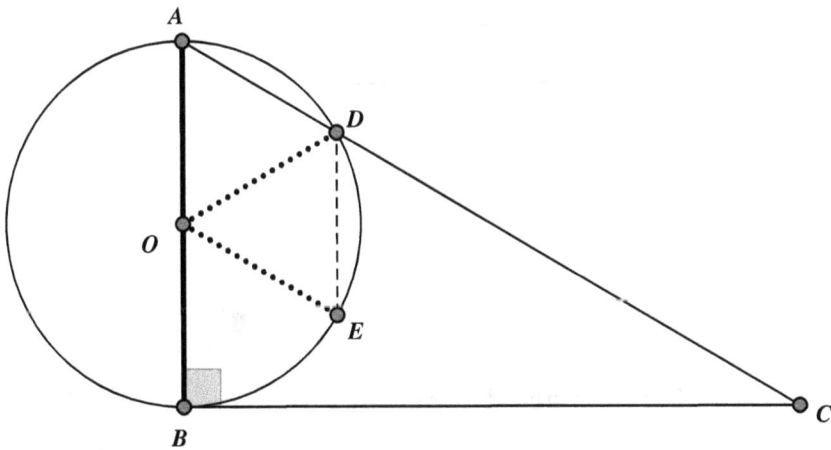

Figure 15-P

## Curiosity 16. A Surprising Perpendicularity

This is immediately clear when we draw radius $OT$, as shown in Figure 16-P. The radius $OT$ is perpendicular to the tangent in $T$. As $\angle BOT$ is an exterior angle of isosceles triangle $AOT$, we have $\angle BAT = \frac{1}{2} \cdot \angle BOT$ in circle $O$. We also have $\angle BAT = \frac{1}{2} \cdot \angle BAC$, and we obtain $\angle BOT = \angle BAC$; therefore, $AC \| OT$. Since $OT$ is perpendicular to the tangent at point $T$, $AC$ must then also be perpendicular to this tangent.

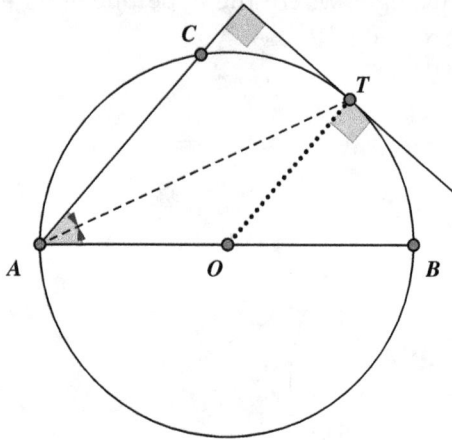

Figure 16-P

## Curiosity 17. Unexpected Concyclic Points

As we know from Curiosity 6, the angle formed by two intersecting chords in a circle is equal to one-half the sum of the intercepted arcs. In Figure 17-P, this gives us $\angle AGC = \frac{1}{2} \cdot (\angle AOC + \angle BOP)$. Noting that point $P$ is the midpoint of arc $AB$, we have $\angle BOP = \angle POA$. Therefore, we have $\angle AGC = \frac{1}{2}(\angle AOC + \angle BOP) = \frac{1}{2}(\angle AOC + \angle POA) = \frac{1}{2}\angle POC = \angle PDC$, which enables us to conclude that quadrilateral $CDHG$ is cyclic.

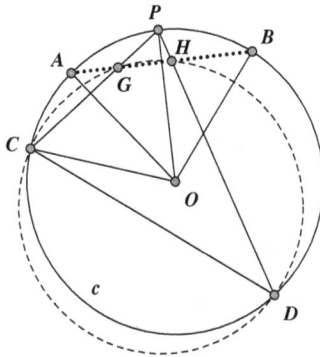

Figure 17-P

## Curiosity 18. Circle Chords Generate a Constant Angle Sum

In Figure 18-P, we first note that $\angle CDA = \frac{1}{2}\widehat{ABC} = \frac{1}{2}\left(360° - \widehat{ADC}\right) = 180° - \frac{1}{2}\widehat{ADC} = 180° - \frac{1}{2}\angle AOC$. Similarly, we can get $\angle BEC = 180° - \frac{1}{2} \cdot \angle COB$. Point $P$ is the midpoint of arc $AB$ containing points $C$ and $D$. We then obtain

$$\angle CDA + \angle BEC = \left(180° - \frac{1}{2} \cdot \angle AOC\right) + \left(180° - \frac{1}{2} \cdot \angle COB\right)$$

$$= 360° - \frac{1}{2} \cdot \left(\angle AOC + \angle COB\right) = 360° - \frac{1}{2} \cdot \angle AOP.$$

We therefore see that the value of $\angle CDA + \angle BEC$ is independent of the locations of these points $C$, $D$, and $E$.

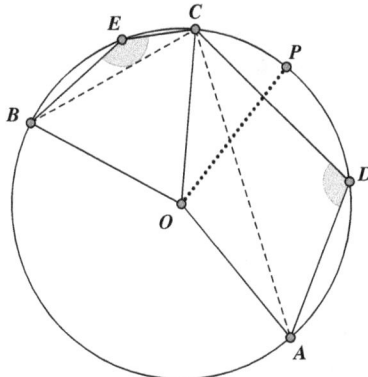

Figure 18-P

### Curiosity 19. A Circle Helps Divide a Line Segment into Two Equal Parts

As we see in Figure 19-P, the two perpendiculars $DE$ and $BF$ are parallel, so that $\frac{EP}{BC} = \frac{AE}{AB}$. Furthermore, we also have $\frac{DP}{CF} = \frac{AP}{AC} = \frac{AE}{AB} = \frac{EP}{BC}$. Now, consider triangle $ABF$, where $OC||AF$. Since $O$ is the midpoint of $AB$, it follows that $C$ must be the midpoint of $BF$, and so $CF = BC$. Substituting in $\frac{DP}{CF} = \frac{EP}{BC}$, we have $DP = EP$.

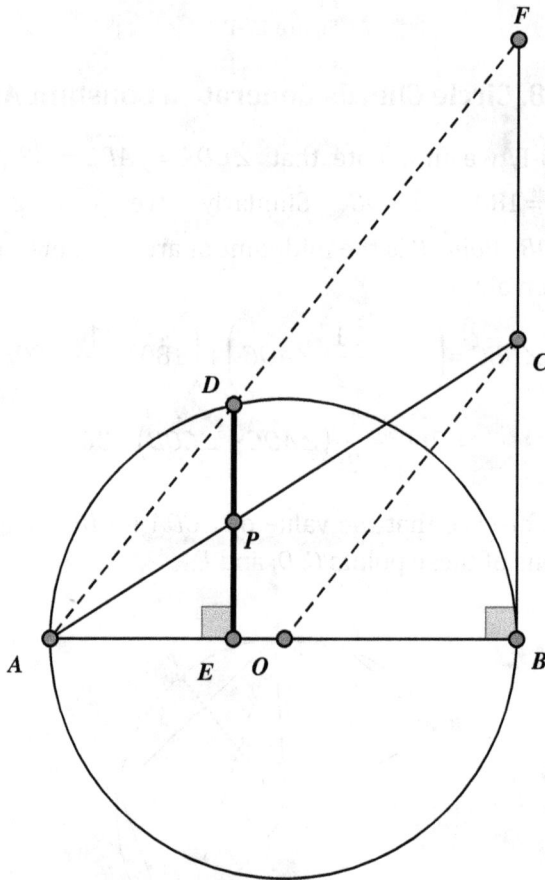

Figure 19-P

## Curiosity 20. The Famous Butterfly Relationship

We begin by drawing the auxiliary lines $FG||AB$ and $MN \perp FG$, as well as the lines $GM$, $GQ$, and $GD$, as shown in Figure 20-P. As $MN \perp AB$, $MN$ contains the center $O$ of the circle. Since $MN$ is also the perpendicular bisector of $FG$, and we have $MF = MG$, it follows that $\triangle MFN \cong \triangle MNG$ and $\angle FMN = \angle NMG$. Furthermore, we have $\angle AMF = \angle GMB$, since they are complements of the equal angles $\angle FMN = \angle NMG$. Since $FG||AB$, we can conclude that $\overparen{AF} = \overparen{GB}$. From Curiosity 6, we know that the angle formed by two intersecting chords is one-half the sum of the intercepted arcs, and we can apply this to obtain $\angle GMB = \angle AMF = \frac{1}{2} \cdot \left( \overparen{AF} + \overparen{BE} \right) = \frac{1}{2}\left( \overparen{GB} + \overparen{BE} \right)$. We know that $\angle EDG = \frac{1}{2} \cdot \overparen{EAG}$, and by addition, $\angle GMB + \angle EDG = \frac{1}{2} \cdot \left( \overparen{GB} + \overparen{BE} + \overparen{EAG} \right) = \frac{1}{2} \cdot 360° = 180°$. Since this means $\angle GMQ + \angle QDG = 180°$, we see that quadrilateral $MGDQ$ is cyclic, and we obtain $\angle QDM = \angle QGM$. We then obtain $\angle QGM = \angle QDM = \angle EDC = \angle EFC = \angle MFP$ (note that $\angle EDC = \angle EFC$, as they are both measured by one-half of arc $FGD$), which enables us to establish that $\triangle MPF \cong \triangle MGQ$ (ASA). Thus, $MP = MQ$.

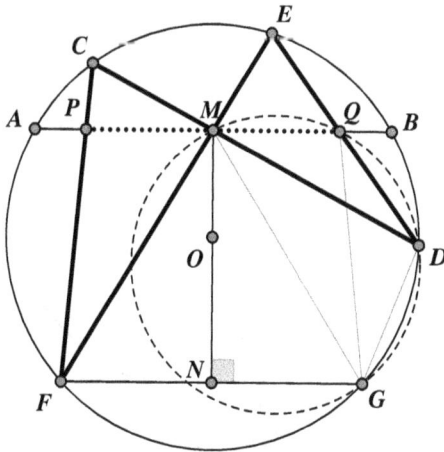

Figure 20-P

## Curiosity 21. A Circle Unexpectedly Produces Parallel Lines

We begin by drawing the line *CB* in Figure 21-P, and since *PB* is tangent to circle *O*, we have two angles whose measures are one-half the measure of arc *BC* with $\angle PBC = \angle BAC$. Since the tangents from *P* to the circle are of equal length $PC = PB$, that triangle *PCB* is isosceles and the bisector of $\angle CPB$ contains point *O*. Since the bisector at vertex *P* is also the altitude of the isosceles triangle *PCB*, *PO* is perpendicular to the chord *BC* and intersects *BC* in its midpoint *F*. In right triangle *PFB*, we have $\angle PBF + \angle FPB = 90°$, and in right triangle *BPO*, we have $\angle BOP + \angle OPB = 90°$. Therefore, we have $\angle PBF = 90° - \angle FPB = 90° - \angle OPB = \angle BOP$, and thus, $\angle BAC = \angle PBC = \angle PBF = \angle BOP$, which means that *AC* is parallel to *OP*.

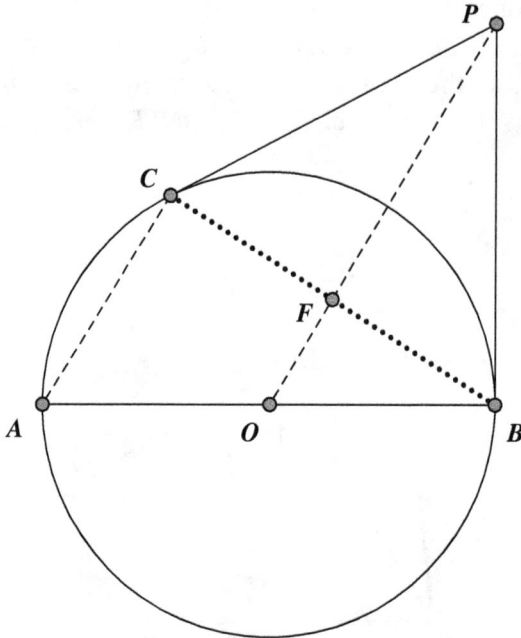

Figure 21-P

## Curiosity 22. The Unexpected Constant Point on a Circle

In Figure 22-P, we are given $CO = DO$ and we add point $R$ on diameter $AB$ such that $PR \perp AOB$. Also, we draw radius $OP$. We apply the Pythagorean theorem to triangle $PCR$ to get $PC^2 = PR^2 + CR^2$, and we apply the Pythagorean theorem to triangle $PRD$ to get $PD^2 = PR^2 + DR^2$. By addition, this gives us $PC^2 + PD^2 = 2PR^2 + CR^2 + DR^2$. Furthermore, applying the Pythagorean theorem to triangle $PRO$, we get $PR^2 = PO^2 - RO^2$. Since $CR = CO - RO$, we have $CR^2 = (CO - RO)^2$. Similarly, since $DR = RO + DO$, we have $DR^2 = (RO + DO)^2$. Now making appropriate substitutions in the earlier equation, we have

$$PC^2 + PD^2 = 2 \cdot (PO^2 - RO^2) + (CO - RO)^2 + (RO + DO)^2$$
$$= 2PO^2 - 2RO^2 + CO^2 - 2 \cdot CO \cdot RO + RO^2 + RO^2 + 2 \cdot RO \cdot DO + DO^2$$
$$= 2PO^2 + CO^2 + DO^2 = 2PO^2 + 2CO^2,$$

since $CO = DO$. In short, we obtain $PC^2 + PD^2 = 2PO^2 + 2CO^2$. Since $PO$ and $DO$ are constant quantities, independent of the location of point $P$ on the circle, the sum $PC^2 + PD^2$ is also a constant.

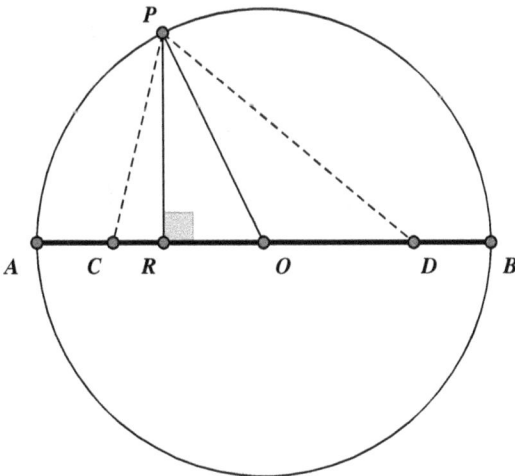

Figure 22-P

## Curiosity 23. Circle Tangents Generate Perpendicular Lines and Proportional Segments

Since the radius to the point of tangency is perpendicular to the tangent line, we have right angles $\angle OAD$, $\angle CBO$, $\angle OEC$, and $\angle DEO$, as shown in Figure 23-P. Furthermore, since $OD$ bisects $\angle EOA$, and $CO$ bisects $\angle BOE$, we have $\angle COD = \angle COE + \angle EOD = \frac{1}{2}\angle BOE + \frac{1}{2}\angle EOA = \frac{1}{2} \cdot 180° = 90°$. We now have $OE$ as the altitude in right triangle $OCD$, so that $\angle COE = 90° = -\angle ECO = 90° - \angle DCO = \angle ODC$. From this, we obtain $\triangle EOC \sim \triangle EDO$, whereupon it follows that $\frac{OE}{CE} = \frac{DE}{OE}$. However, since $EC = BC$, and $AD = DE$, we then have $\frac{OE}{BC} = \frac{AD}{OE}$.

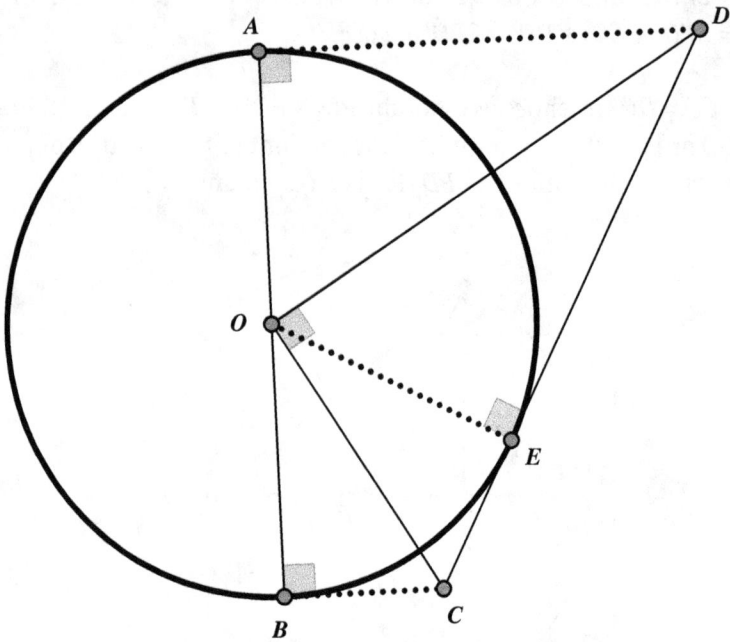

Figure 23-P

## Curiosity 24. An Unexpected Circle Diameter

In Figure 24-P, we draw lines $AC$, $CX$, and $CY$, and then notice that $\angle YCA = \frac{1}{2}\widehat{AY} = \angle YDA$ and $\angle ACX = \frac{1}{2}\widehat{AX} = \angle ABX$. This enables us to get $\angle YCX = \angle YCA + \angle ACX = \angle YDA + \angle ABX$. Since $DY$ and $BX$ are the bisectors of angles $\angle CDA$ and $\angle ABC$, respectively, this gives us $\angle YCX = \angle YDA + \angle ABX = \frac{1}{2}\angle CDA + \frac{1}{2}\angle ABC = \frac{1}{2}(\angle CDA + \angle ABC)$. We know that quadrilateral $ABCD$ is cyclic, and the opposite angles are supplementary, so that $\angle CDA + \angle ABC = 180°$. Therefore, $\angle YCX = \frac{1}{2} \cdot (\angle CDA + \angle ABC) = \frac{1}{2} \cdot 180° = 90°$, and $XY$ is thus the diameter of the given circle.

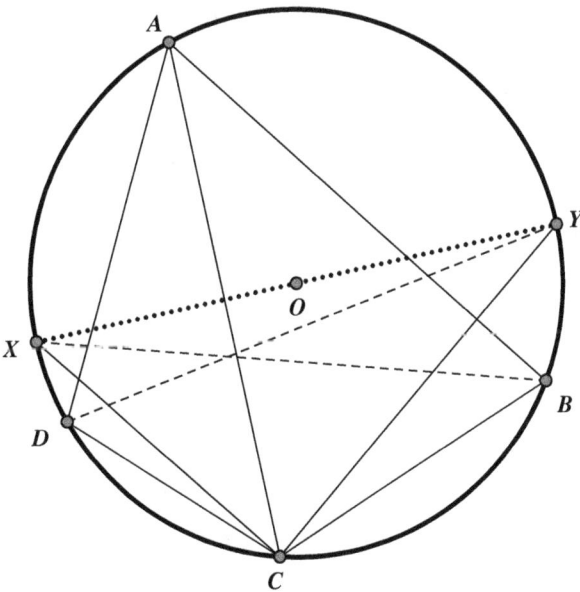

Figure 24-P

### Curiosity 25. An Unexpected Rectangle in a Circle

In Figure 25-P, we first draw lines *CE* and *AE*. Since we have *FP||BD*, we also have corresponding angles $\angle PFE = \angle DBE$. In circle *O*, we have $\angle PFE = \angle DBE = \angle DCE = \angle PCE$, which establishes that quadrilateral *CPEF* is cyclic, and its opposite angles are supplementary. Thus, we have $\angle CFE = 90°$, so that $\angle EPD = \angle EPC = 90°$. Similarly, in circle *O* we also have $\angle QAE = \angle DAE = \angle DBE = \angle PFE = \angle QFE$, and quadrilateral *AFQE* is also cyclic. Since $\angle EFA = 90°$, *AE* is the diameter of the circumcircle of *AFQE*, and we obtain $\angle EQA = 90°$. Therefore, we have shown that $\angle EQA = \angle DQE = \angle EPD = 90°$, and since $\angle BDQ = 180° - \angle ADC = 180° - 90° = 90°$, we have three right angles in quadrilateral *EQDB*, establishing it as a rectangle.

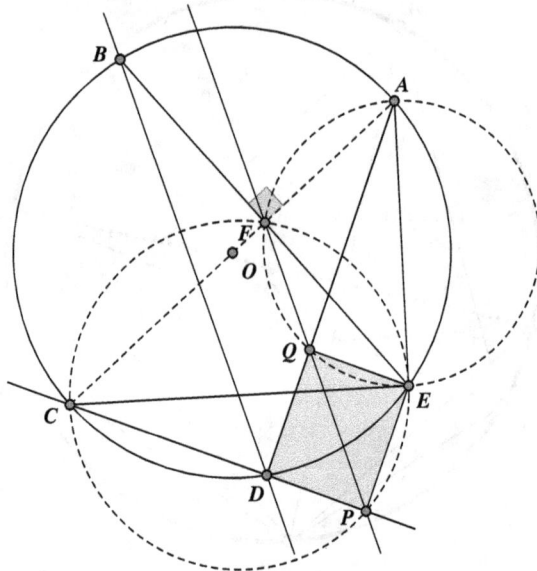

Figure 25-P

### Curiosity 26. Parallel Angle Bisectors

We begin by drawing points *X* and *Y* in Figure 26-P, such that ray *EX* is the bisector of $\angle DEA$, and ray *FY* is the bisector of $\angle DFA$. Furthermore, we label the intersection point *G* of lines *FY* and *BD*. In triangle *ABE*, we note that $\angle DEA = \angle BAE + \angle EBA = \angle FAC + \angle DBA$, and since $EX$ is the bisector of $\angle DEA$, we have $\angle DEX = \frac{1}{2}(\angle FAC + \angle DBA)$. Now consider triangle $BFD$. Since angles measured by the same arc $BC$ are equal, we get $\angle BDF = \angle BDC = \angle BAC = \angle FAC$. Furthermore, since $\angle FBD = 180° - \angle DBA$, we obtain $\angle DFB = 180° - \angle BDF - \angle FBD = 180° - \angle FAC - (180° - \angle DBA) = \angle DBA - \angle FAC$. Because $FY$ is the bisector of $\angle DFA$, we have $\angle DFY = \frac{1}{2}(\angle DBA - \angle FAC)$. Considering triangle $GFD$, we get the following:

$$\angle DGY = \angle DFG + \angle GDF$$
$$= \angle DFY + \angle BDF$$
$$= \frac{1}{2}(\angle DBA - \angle FAC) + \angle FAC$$
$$= \frac{1}{2}(\angle FAC + \angle DBA).$$

Since both $\angle DEX$ and $\angle DGY$ are equal to $\frac{1}{2}(\angle FAC + \angle DBA)$, the angle bisectors $EX$ and $FY$ are parallel.

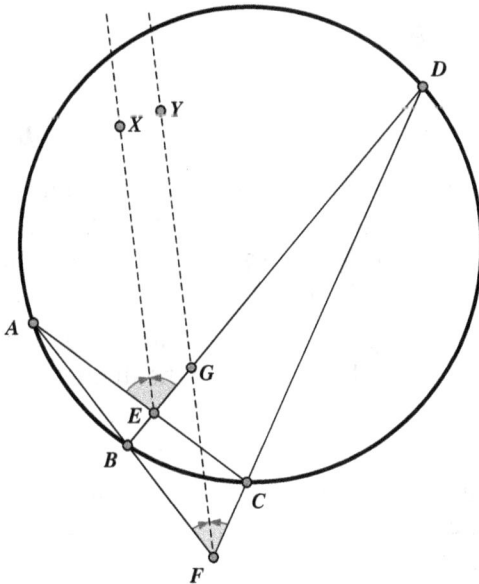

Figure 26-P

## Curiosity 27. Noteworthy Perpendiculars Generated by a Semicircle

In Figure 27-P, point $O$ is the center of the semicircle's diameter. Since $OP = OQ$, triangle $OQP$ is isosceles, and since $M$ is the midpoint of its base $PQ$, we have $OM \perp PQ$. We now note that $M$ is the midpoint of $AC$ and $O$ is the midpoint of $AB$. This gives us $BC \| OM$, and we therefore have $BC \perp PQ$.

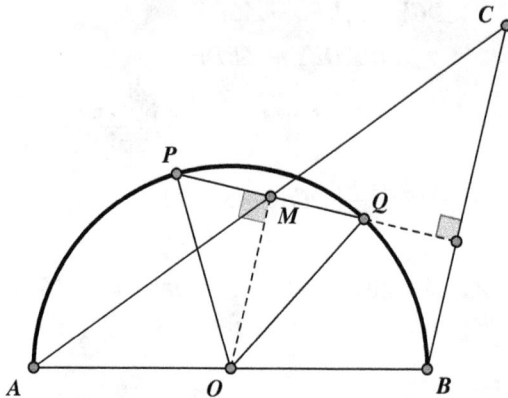

Figure 27-P

## Curiosity 28. An Unexpected Angle Bisector in a Semicircle

In Figure 28-P, we draw lines $AP$ and $BQ$. Since point $P$ is on the semicircle with diameter $AB$, we have $\angle APB = 90°$. Also $\angle RSA = 90°$, so that quadrilateral $ASRP$ is cyclic and $AR$ is the diameter of its circumcircle. We therefore have $\angle RSP = \angle RAP$, since they are both measured by one-half arc $PR$. Similarly, we also have $\angle AQB = 90°$. Since $\angle BSR = 90°$, quadrilateral $BQRS$ is cyclic with $BR$ as the diameter of its circumcircle, and we have $\angle QSR = \angle QBR$. We have $\angle QAP = \angle QBP$ in the given semicircle, as they are both one half the measure of arc $PQ$. From the above information we then obtain $\angle RSP = \angle RAP = \angle QAP = \angle QBP = \angle QSR$, and thus, $\angle RSP = \angle QSR$, which establishes that $RS$ is the angle bisector of $\angle QSP$.

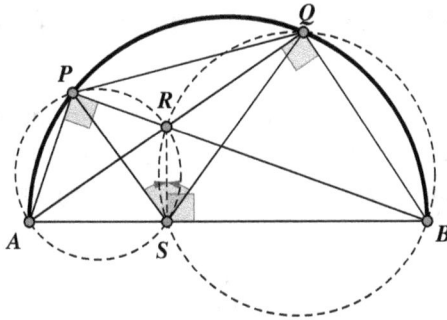

Figure 28-P

## Curiosity 29. Tangent Circles on a Semicircle

We locate the midpoint $O$ of $AB$ and draw line $OP$, as shown in Figure 29-P. Point $O$ is the center of the semicircle and $OP$ is a radius of the semicircle. Since $OA$ and $OP$ are both radii, triangle $OPA$ is isosceles, and we have $\angle APO = \angle OAP$. Furthermore, in circle $O$, we also have angles measured by arc $PCB$, whereupon $\angle OAP = \angle PAB = \angle BDP = \angle FDP$. Since $FP$ is a chord in the circumcircle of triangle $PDF$, and since equal angles $FDP$ and $FPO$ are each one-half the measure of arc $PF$, we can conclude that $OP$ is tangent to this circle at point $P$. In an analogous way, we can also show that $OP$ is tangent to the circumcircle of triangle $PEC$ at point $P$, and these two circles are therefore also tangent to each other at point $P$.

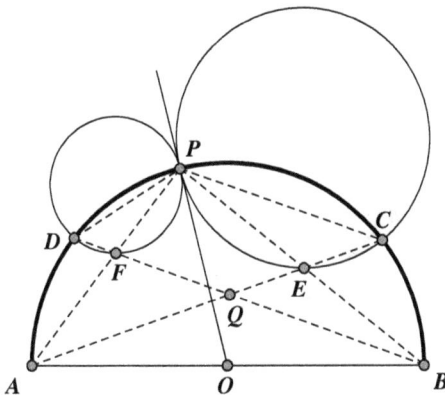

Figure 29-P

## Curiosity 30. A Point on the Circumcircle of an Isosceles Triangle

By applying Ptolemy's theorem (see Toolbox) to quadrilateral *ABPC*, as shown in Figure 30-P, we get $PA \cdot BC = PB \cdot AC + PC \cdot AB$. Since $AB = AC$, we therefore have $PA \cdot BC = AC \, (PB + PC)$, and thus, $\frac{AC}{BC} = \frac{PA}{PB+PC}$.

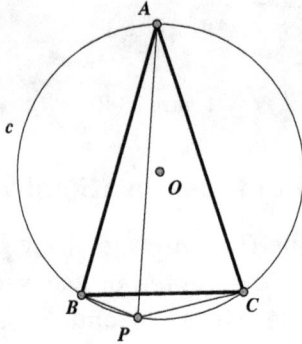

Figure 30-P

## Curiosity 31. A Point on the Circumcircle of a Square

In Figure 31-P, we have added the diagonals *AC* and *BD* of square *ABCD*. We can apply the result from Curiosity 30 to isosceles triangle *ABD*, in which we have $AB = AD$, to obtain $\frac{AD}{BD} = \frac{PA}{PB+PD}$. Repeating this with isosceles triangle *ADC*, we get $\frac{DC}{AC} = \frac{PD}{PA+PC}$. Since $AD = DC$, and $BD = AC$, we therefore get $\frac{PA}{PB+PD} = \frac{PD}{PA+PC}$, or $\frac{PA+PC}{PB+PD} = \frac{PD}{PA}$.

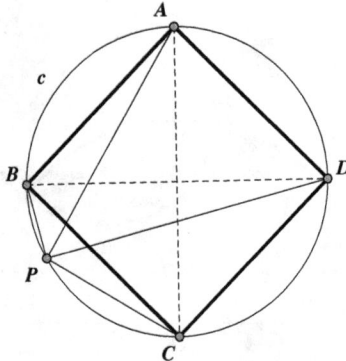

Figure 31-P

## Curiosity 32. A Point on the Circumcircle of a Regular Pentagon

In Figure 32-P, we have first added the diagonals of pentagon *ABCDE*. As with the previous two Curiosities, it will once again prove to be useful to apply Ptolemy's theorem to several quadrilaterals in the figure. By first considering quadrilateral *ABPC*, we get

$$PA \cdot BC = AB \cdot PC + PB \cdot AC. \tag{I}$$

Similarly, for quadrilateral *BPCD*, we get

$$PD \cdot BC = PB \cdot CD + PC \cdot BD. \tag{II}$$

Since $BA = CD$ and $AC = BD$, adding (I) and (II) gives us

$$BC(PA + PD) = BA(PB + PC) + AC(PB + PC). \tag{III}$$

Since $\triangle BCE$ is isosceles, we know from Curiosity 30 that

$$\frac{CE}{BC} = \frac{PE}{PB + PC}, \text{ or } \frac{PE \cdot BC}{PB + PC} = CE = AC. \tag{IV}$$

Substituting (IV) into (III) then gives us

$$BC \cdot (PA + PD) = BA \cdot (PB + PC) + \frac{PE \cdot BC}{PB + PC} \cdot (PB + PC),$$

and since $BC = BA$, this reduces to $PA + PD = PB + PC + PE$.

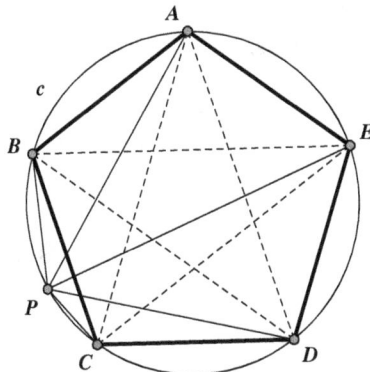

Figure 32-P

## Curiosity 33. A Point on the Circumcircle of a Regular Hexagon

This is an immediate consequence of Curiosity 30. As we see in Figure 33-P, since *ABCDEF* is a regular hexagon, triangles *ACE* and *BDF* are both equilateral. Considering triangle *ACE*, Curiosity 30 tells us that $\frac{CE}{AC} = \frac{PE}{PA+PC}$, and since $CE = AC$, this is equivalent to $PE = PA + PC$. In the same way, equilateral triangle *BDF* gives us $PF = PB + PD$, and we add these two equalities to get $PE + PF = PA + PB + PC + PD$.

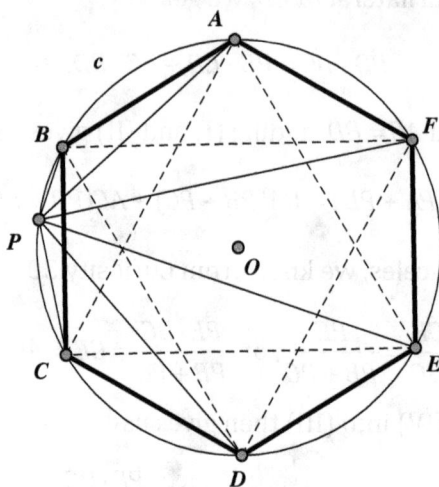

Figure 33-P

## Curiosity 34. The Tangent to the Circumscribed Circle of an Equilateral Triangle

We begin the proof by extending altitude *AM* to meet the circumscribed circle at point *N*, as shown in Figure 34-P. Then we draw auxiliary lines *OP*, *MK*, and *NL*, each perpendicular to the tangent *DF* of circle *O*. Because triangle *ABC* is equilateral, point *O* is the intersection of the altitudes as well as the medians and is, therefore, the trisection point of altitude $AM$. Since $OM$ is half the length of $AO$, it is also half the length of $ON$, and $M$ is the midpoint of $ON$. This makes $MK$ the midline of trapezoid $BCFE$, and we have $BE + CF = 2MK$. Since $MK$ is also the midline of trapezoid $NOPL$, we also obtain $OP + NL = 2MK$, and this results in $BE + CF = OP + NL$. Analogously, trapezoid $ADLN$ gives us $AD + NL = 2OP$, which is equivalent to $AD + NL + OP = 3OP$. Substituting $BE + CF$ for $OP + NL$ in this equation gives us $AD + BE + CF = 3OP$, and since $OP = \frac{2}{3}AM$, we obtain $AD + BE + CF = 3OP = 3 \cdot \frac{2}{3} \cdot AM = 2AM$.

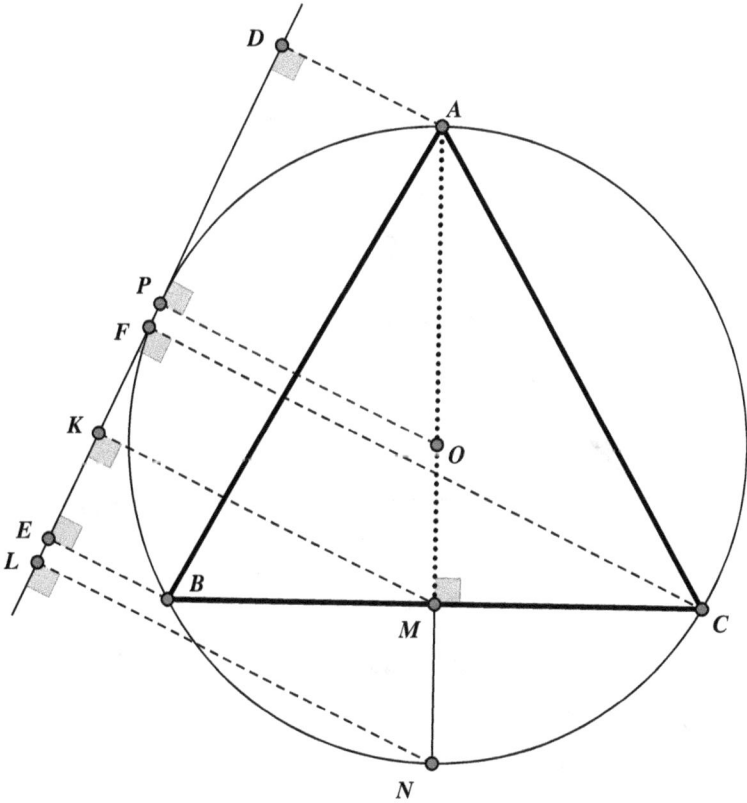

Figure 34-P

## Curiosity 35. Surprising Concyclic Points

As a first step, we draw lines *OD*, *OE*, and *EF* in Figure 35-P, with point *O* being the center of the circle. We note that, since *AB*||*ED*, we have $\angle DEC = \angle BAC$, and since the circle is tangent to *AC* at point *E*, we have $\angle OED = \angle OEC - \angle DEC = 90° - \angle BAC$. For isosceles triangle *ODE*, we have $\angle DOE = 180° - 2\angle ODE = 180° - 2(90° - \angle BAC) = 2\angle BAC$. Since point *F* lies on the circle, $\angle DFE = \frac{1}{2} \cdot \angle DOE = \angle BAC$, and thus $\angle EFB = 180° - \angle DFE = 180° - \angle BAC$. We see that quadrilateral *ABFE* is cyclic, since opposite angles are supplementary, that is, $\angle EFB = 180° - \angle BAC$.

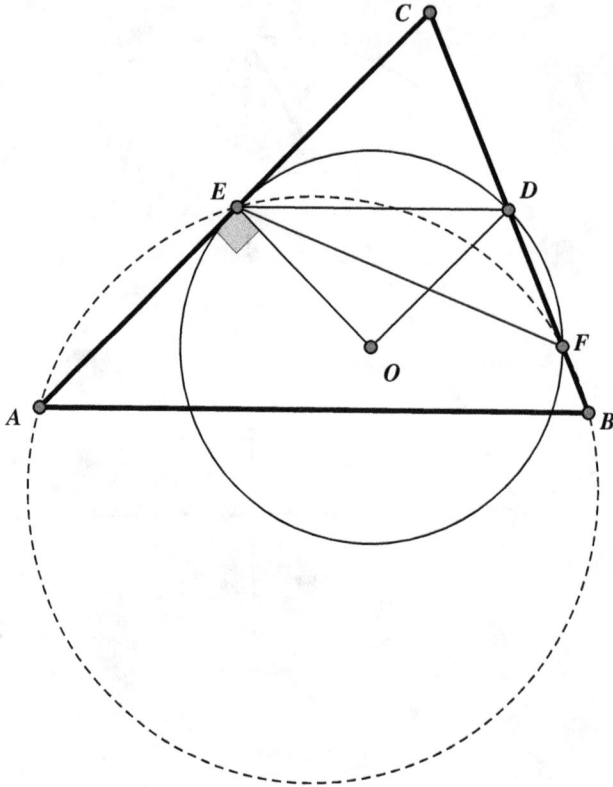

Figure 35-P

## Curiosity 36. The Midpoints of the Arcs of a Circle Produce Unexpected Parallels

In Figure 36-P, we first note that $\angle BAP = \angle PAC$, since $P$ is the midpoint of arc $BC$, so that $AP$ is the angle bisector of $\angle BAC$. Analogously, $BQ$ and $CR$ are also angle bisectors of $\angle CBA$ and $\angle ACB$, respectively. Thus, line segments $AP$, $BQ$, and $CR$ are the angle bisectors of triangle $ABC$, and their point of intersection $I$ is therefore the incenter of $ABC$. Also, we note that angles inscribed in arc $PC$ are equal, yielding $\angle PAC = \angle PRC$. We therefore obtain $\angle DAI = \angle BAP = \angle PAC = \angle PRC = \angle DRI$, and quadrilateral $ARDI$ is then a cyclic quadrilateral. In the circumcircle of quadrilateral $ARDI$, we obtain $\angle IRA = \angle IDA$, and in the original circle we have $\angle CRA = \angle CBA$. Summing up, we now have $\angle IDA = \angle IRA = \angle CRA = \angle CBA$, and $ID$ is therefore parallel to $BC$. Since we can make an analogous argument for quadrilateral $AIEQ$, yielding $IE||BC$, we see that points $D$, $I$, and $E$ are collinear, with $DIE||BC$.

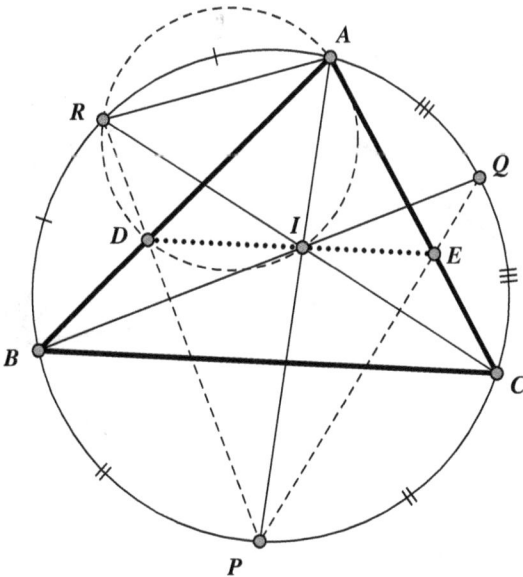

Figure 36-P

## Curiosity 37. The Circumscribed Circle and the Altitudes of a Triangle Reveal an Unusual Equality

In Figure 37-P, we first note that the points $B$, $C$, $E$, and $F$ are concyclic, since $\angle BEC = 90° = \angle BFC$. Because the opposite angles of a cyclic quadrilateral are supplementary, we have $\angle ECB = 180° - \angle BFE$, and since $\angle EFA = 180° - \angle BFE$, we obtain $\angle ECB = \angle EFA$. Furthermore, in triangle $APF$, we have $\angle QPA = \angle FPA = \angle QFA - \angle PAF = \angle QFA - \angle PAB$. We now note that $\angle PAB = \angle PQB$, since they are both measured by one-half arc $PB$, and $\angle AQB = \angle ACB$, since they are both measured by one-half arc $AB$. We then have $\angle AQP = \angle AQB - \angle PQB = \angle ACB - \angle PAB = \angle ECB - \angle PAF = \angle EFA - \angle PAF = \angle EPA$. Since this gives us $\angle AQP = \angle FPA = \angle QPA$, triangle $APQ$ is isosceles, and we have $AP = AQ$.

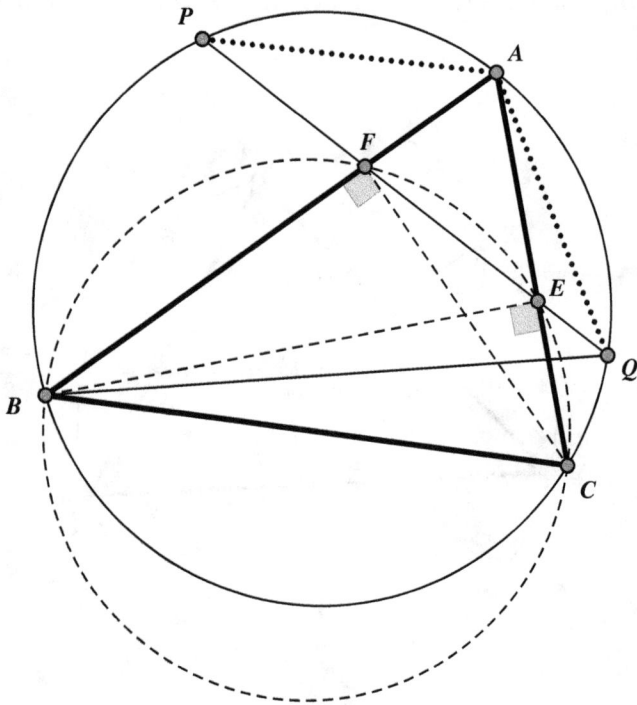

Figure 37-P

## Curiosity 38. The Unexpected Relationship Between the Midpoint of the Side of an Acute Triangle, its Circumcenter and its Orthocenter

In Figure 38-P, *AD*, *BE*, and *CF* are the altitudes of triangle *ABC*, with point *D* on *BC*, point *E* on *CA*, and point *F* on *AB*. Point *P* is the diametrically opposite point to *B* on the circumcircle of $\triangle ABC$. Since *BP* is a diameter of circle *O*, we have $PC \perp BC$ and $PA \perp AB$. Since *AD* and *CF* are altitudes in triangle *ABC*, we also have $AH \perp BC$ and $CH \perp AB$, which gives us $AH \| PC$ and $CH \| PA$. We see that *AHCP* is a parallelogram, and thus, $AH = PC$. The perpendicular from point *O* to *BC* intersects *BC* at its midpoint *M*. In triangle *PBC*, line *OM* thus joins the midpoints of sides *PB* and *BC* and is therefore one-half the length of *PC*, so that $OM = \frac{1}{2}AH$.

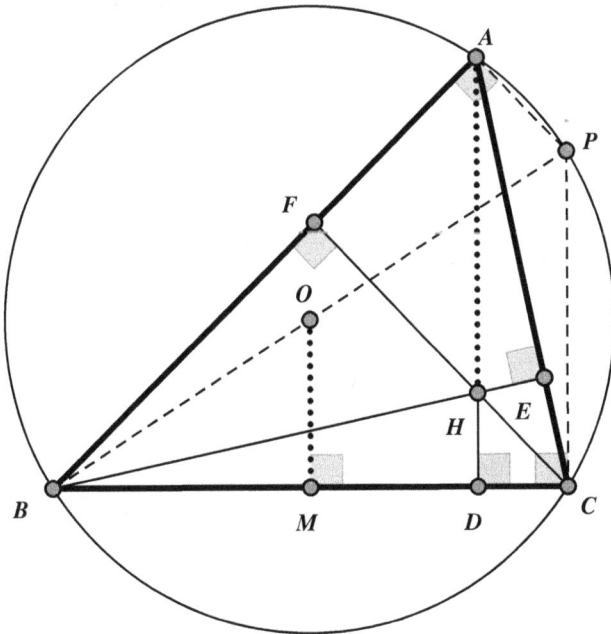

Figure 38-P

## Curiosity 39. An Unexpected Concurrency with the Circumscribed Circle of the Triangle

In Figure 39-P, *COQ* is a diameter in the circumcircle of triangle *ABC*. Our initial goal here is to prove that quadrilateral *HAQB* is a parallelogram. Since *CQ* is a diameter of circle *O*, we have $\angle BQA = 180° - \angle ACB$. Also, since *BH* is an altitude in triangle *ABC* and the angles subtended on the arc of the circle are equal, we have

$$\begin{aligned} \angle HBQ &= \angle HBA + \angle ABQ \\ &= (90° - \angle BAC) + \angle ACQ \\ &= (90° - \angle BAC) + (90° - \angle CQA) \\ &= (90° - \angle BAC) + (90° - \angle CBA) \\ &= 180° - \angle BAC - \angle CBA \\ &= \angle ACB. \end{aligned}$$

We can determine that $\angle QAH = \angle ACB$ in an analogous way. With $\angle HBQ = \angle QAH = \angle ACB$, and since the opposite angles of the cyclic quadrilateral *AQBC* are supplementary, we have $\angle BQA = 180° - \angle ACB$. The angle *HBQ* is therefore supplementary to angle *BQA*, and thus, $AQ \| HB$ so that quadrilateral *HAQB* is a parallelogram. This means that its diagonals *AB* and *HQ* bisect each other in the midpoint *N* of *AB*. Point *Q* is therefore also the point *P* in which *HN* intersects circle *O*, which is what we set out to prove.

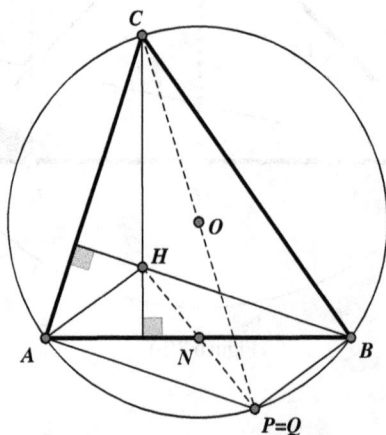

Figure 39-P

## Curiosity 40. An Unexpected Angle Bisector Hidden Within the Circumscribed Circle

In Figure 40-P, we draw lines $OM$, $ON$, and $OB$. Since $OB$ is a radius of the circle and $NB$ is a tangent to circle $O$, they are perpendicular. Since the two tangents $NA$ and $NB$ are equal, $ON$ is the perpendicular bisector of base $AB$ of triangle $NBA$. We also have $\angle NOB = \frac{1}{2}\overset{\frown}{ADB} = \angle ABN$, and $\angle ABN = \angle ACB$, since both are one-half the measure of $\overset{\frown}{ADB}$, so that we obtain $\angle NOB = \angle ABN = \angle ACB$. This means that right triangles $BON$ and $PBC$ are also similar, and thus, $\frac{BN}{PB} = \frac{BO}{PC}$ or $PC = \frac{BO \cdot PB}{BN}$. Similarly, we can show that $\triangle BMO \sim \triangle PAB$, giving us $\frac{BM}{PB} = \frac{BO}{PA}$ or $PA = \frac{BO \cdot PB}{PM}$. This gives us

$$\frac{PA}{PC} = \frac{\dfrac{BO \cdot PB}{BM}}{\dfrac{BO \cdot PB}{BN}} = \frac{BN}{PB}.$$

Since two tangents from an external point to a circle are equal, we have $AN = BN$ and $BM = CM$, and thus obtain $\frac{PA}{PC} = \frac{BN}{BM} = \frac{AN}{CM}$. Therefore, we have $\triangle ANP \sim \triangle CPM$ and $\angle APN = \angle MPC$. We therefore obtain $\angle NPB = \angle APB - \angle APN = 90° - \angle APN = 90° - \angle MPC = \angle BPC - \angle MPC = \angle BPM$, and $PB$ bisects $\angle NPM$.

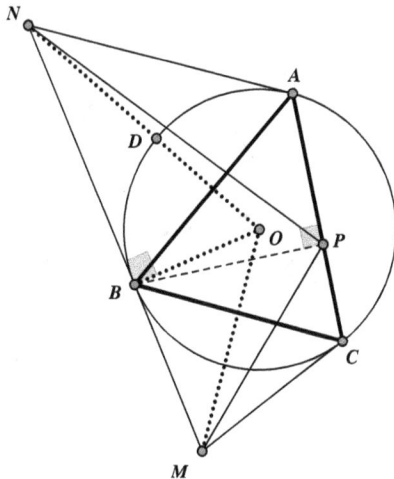

Figure 40-P

## Curiosity 41. Surprising Concyclic Points Generated by a Triangle and its Circumscribed Circle

In Figure 41-P, we first add line $QR$. Since $DE \| AB$, we have $\angle PED = \angle RBA$, and since the angles measured by the common arc $RA$ are equal, we have $\angle RBA = \angle RQA$. Therefore, we obtain $\angle PED = \angle RQA$. Since $\angle DER$ is supplementary to $\angle PED$, it follows that $\angle DER$ is also supplementary to $\angle RQD$, which makes points $D, Q, R$, and $E$ concyclic, since the opposite angles of quadrilateral $DQRE$ are supplementary. As lines $PQ$ and $PR$ are secants to the circumcircle of cyclic quadrilateral $DQRE$, we get $PE \cdot PR = PD \cdot PQ$. Analogously, we can show that the points $D, F, S$, and $Q$ are also concyclic, with $PF \cdot PS = PD \cdot PQ$. Therefore, we have $PE \cdot PR = PF \cdot PS$, which allows us to conclude that the points $S, F, E$, and $R$ all lie on the same circle.

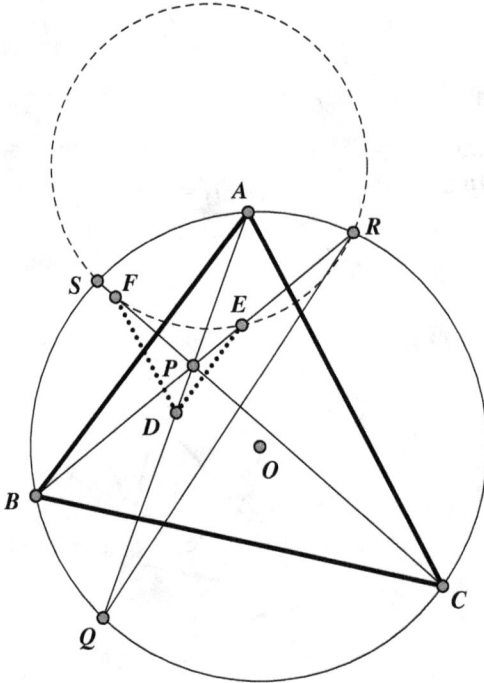

Figure 41-P

## Curiosity 42. Unexpected Concyclic Points

We begin by letting point $N$ be the midpoint of $OA$, as shown in Figure 42-P. Since $\angle ENA$ and $\angle EDA$ are right angles, we know that the points $A$, $N$, $D$, and $E$ are on the circle with diameter $AE$. In triangle $AOH$, line $MN$ is parallel to $AH$, since it joins the midpoints of two sides of the triangle. We also know that when $AD$ is extended to intersect the circumscribed circle of triangle $ABC$ at point $K$, we have $HD = KD$. Therefore, in triangle $OKH$, the line $MD$, which joins the midpoints of sides $MK$ and $MO$, is parallel to $OK$. Also, in triangle $OKH$ we have $\angle HDM = \angle HKO$, and since triangle $AOK$ is isosceles, we therefore have $\angle ADM = \angle HDM = \angle HKO = \angle AKO = \angle KAO = \angle DAN$. We see that quadrilateral $ANMD$ is an isosceles trapezoid, and the circumcircle of triangle $ANE$ thus contains point $M$, which is what we sought to prove.

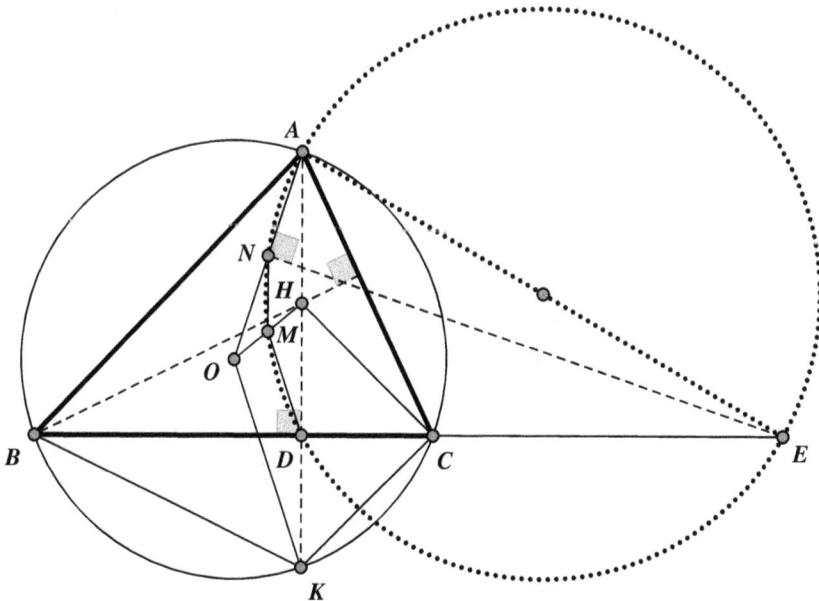

Figure 42-P

## Curiosity 43. The Orthocenter and Two Vertices of a Triangle Generate a Circle Equal to the Circumcircle

In Figure 43-P, we extend altitude $AH$ to meet side $BC$ at point $E$ and the circumscribed circle at point $D$. In cyclic quadrilateral $ABDC$, the opposite angles are supplementary so that $\angle CDB = 180° - \angle BAC$. Furthermore, with $S$ and $T$ as the feet of the altitudes on sides $CA$ and $AB$, respectively, triangles $BCS$ and $BCT$ are right triangles, so that we have

$$\begin{aligned}
\angle BHC &= 180° - \angle CBH - \angle HCB \\
&= 180° - \angle CBS - \angle TCB \\
&= 180° - (90° - \angle SCB) - (90° - \angle CBT) \\
&= \angle ACB + \angle CBA \\
&= 180° - \angle BAC.
\end{aligned}$$

Thus, we have $\angle BHC = \angle CDB$, and since $BC$ is the common chord in circles $O$ and $Q$, and the angles subtended on this common chord in both circles are equal, the circles $O$ and $Q$ are of equal size.

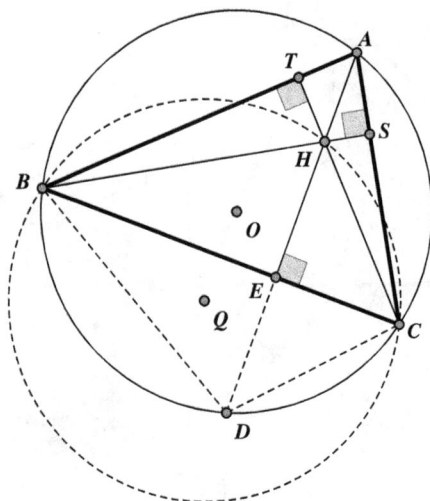

Figure 43-P

## Curiosity 44. The Appearance of an Unexpected Isosceles Triangle

In Figure 44-P, since lines $AB$ and $PQ$ are parallel, we have $\angle CBA = \angle CQP$. Since $PA$ and $PC$ are tangents to circle $O$ at the endpoints of chord $AC$, we have $\angle CBA = \angle CAP = \angle PCA$. This gives us $\angle CQP = \angle CAP$ and quadrilateral $AQCP$ is therefore cyclic, since both angles are measured by one-half arc $AP$ of the circumscribed circle of quadrilateral $AQCP$. In this quadrilateral, we have $\angle PCA = \angle PQA$, and we, thus, have $\angle CQA = \angle CQP + \angle PQA = \angle CBA + \angle PCA = 2\angle CBA$. In triangle $ABQ$, it follows that $\angle BAQ = \angle CQA - \angle QBA = 2\angle CBA - \angle QBA = 2\angle QBA = 2\angle QBA - \angle QBA = \angle QBA$. Since $\angle BAQ = \angle QBA$, triangle $ABQ$ is isosceles, with $AQ = BQ$.

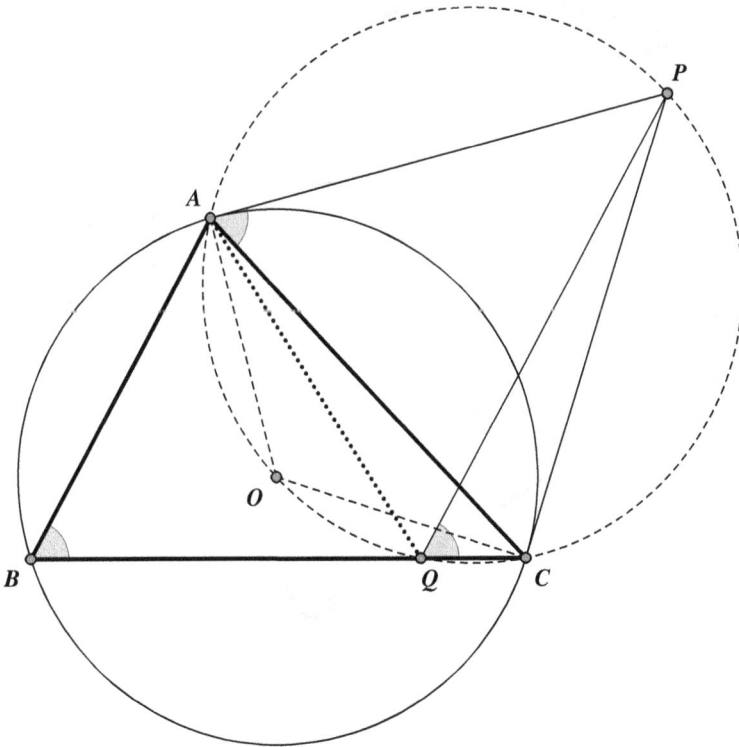

Figure 44-P

## Curiosity 45. Perplexing Parallels from Concyclic Points

In Figure 45-P, since $AB = AE$, we notice that $\angle EBA = \angle ECA = \angle EDA = \angle ACB = \angle ADB = \angle AEB$ because these angles are all measured by equal arcs of circle $O$. From this, we obtain $\angle XDY = \angle ADB = \angle ECA = \angle XCY$, so that we have cyclic quadrilateral $CDXY$.

In the original circle, we have $\angle DBE = \angle DCE$, and in the circumcircle of quadrilateral $CDXY$, we have $\angle DCE = \angle DCX = \angle DYX$. This yields $\angle DBE = \angle DCE = \angle DYX$, and thus, $BE \| XY$.

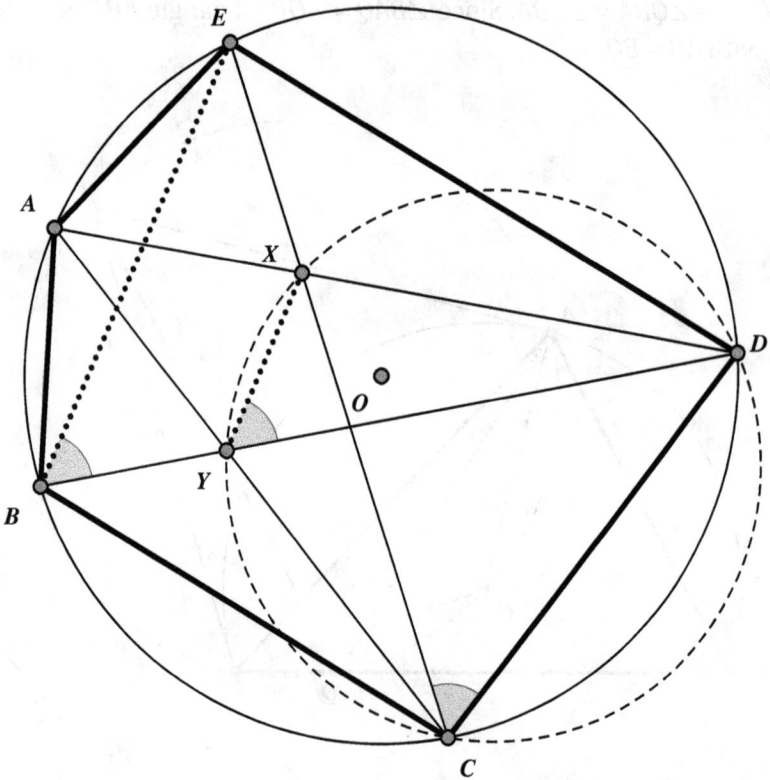

Figure 45-P

## Curiosity 46. Tangents to the Circumcircle of a Triangle

In Figure 46-P, the tangent to the circle at point $A$ is perpendicular to radius $OA$. If we can show that $OA$ is also perpendicular to $EF$, we will have shown that the tangent in $A$ is parallel to $EF$. In order to do this, we first notice that quadrilateral $AFHE$ is cyclic, since $\angle AEH = \angle HFA = 90°$. In right triangle $ABD$, we have $\angle BAD = 90° - \angle DAB = 90° - \angle CBA$, and from the cyclic quadrilateral $AFHE$ we obtain $\angle FEH = \angle FAH = \angle BAD = 90° - \angle CBA$. Considering isosceles triangle $OCA$, we have $\angle OAC = \frac{1}{2}(180° - 2\angle CBA) = 90° - \angle CBA$, and we see that $\angle OPC = \angle FEH = 90° - \angle CBA$. Point $Q$ is the intersection of $OA$ and $EF$, and point $P$ is the intersection of $OA$ and $BE$. In right triangle $AQE$, we have $\angle EPA = 90° - \angle PAE = 90° - \angle OAC = 90° - (90° - \angle CBA) = \angle CBA$, and in triangle $PEQ$, we have $\angle PQE = 180° - \angle EPQ - \angle QEP = 180° - \angle EPA - \angle FEB = 180° - \angle CBA - (90° - \angle CBA) = 90°$. We see that $\angle PQE = 90°$, which gives us $OA \perp EF$, and since we know that the tangent at $A$ is also perpendicular to $OA$, the tangent at $A$ is parallel to $EF$. Similarly, we also obtain that the tangent at $B$ is parallel to $FD$ and the tangent at $C$ is parallel to $DE$.

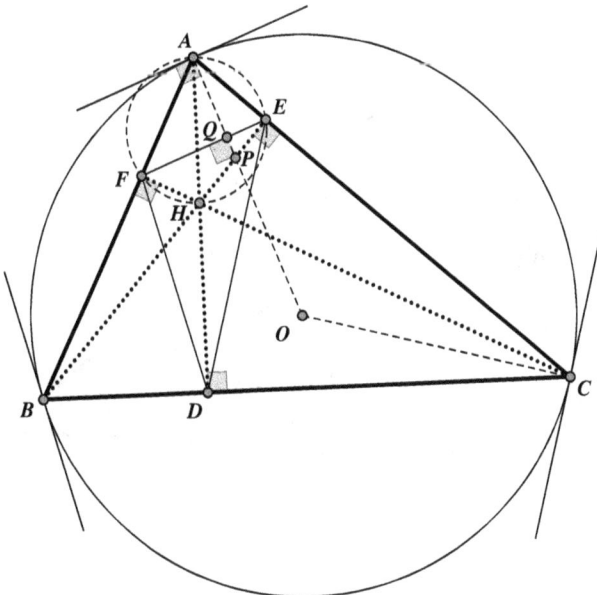

Figure 46-P

## Curiosity 47. Tangents to the Nine-Point Circle

In Figure 47-P, the radius *RK* of the nine-point circle is perpendicular to the tangent at point *K*. We now need to show that $RK \perp EF$. To do this, we consider quadrilateral *KMXY*. Since points *K* and *M* are the midpoints of *CB* and *CA*, respectively, of the sides of triangle *ABC*, *KM* is parallel to *BA* and half as long. Also, since points *Y* and *X* are the midpoints of *HB* and *HA*, respectively, of triangle *HAB*, *YX* is also parallel to *BA* and half as long. Therefore, quadrilateral *KMXY* is a parallelogram because a pair of its opposite sides are equal and parallel. Also, since points *M* and *X* are the midpoints of *AC* and *AH*, respectively, *MX* is parallel to altitude *CH*. We now have *MX||CH*, *XY||AB*, and *CH* ⊥ *AB*, which give us *MX* ⊥ *XY*. Quadrilateral *KMXY* is therefore a rectangle. The diagonal *XK* of this rectangle is a diameter of its circumcircle, and the circle center *R* is the midpoint of *XK*. In the proof of Curiosity 46, we noted that quadrilateral *AFHE* is cyclic, since ∠*HFA* = ∠*AEH* = 90°. The midpoint *X* of diameter *AH* of its circumcircle is equidistant from points *E* and *F*, and we have *XE* = *XF*. Since *EF* is a chord of circle *R*, the altitude of the isosceles triangle is then line *XR*, and line *XRK* is perpendicular to *EF*. Since the tangent to circle *R* at point *K* is also perpendicular to *XRK*, this tangent is parallel to *EF*.

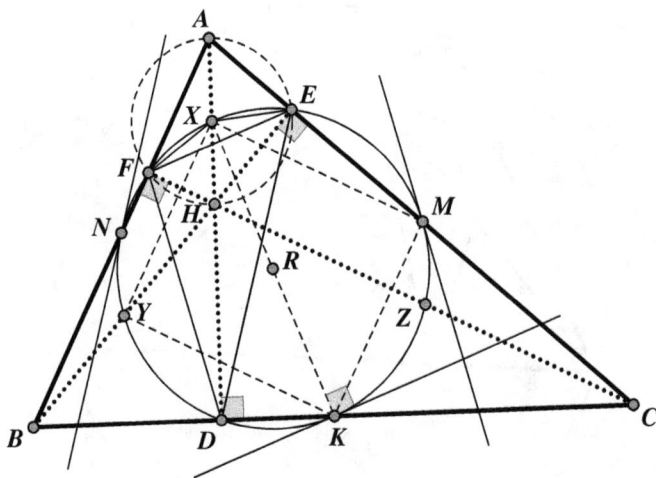

Figure 47-P

This can be repeated for the other two tangent lines to show that the tangent at *M* is parallel to *FD* and the tangent at *N* is parallel to *DE*. Combining this result with Curiosity 46, we see that the tangents of the circumcircle of *ABC* at its vertices are also parallel to the tangents of the Nine-Point Circle at the midpoints of the opposite sides.

## Curiosity 48. An Adventure with the Inscribed Circle of a Triangle and a Special Circumcircle

The point of intersection of the angle bisectors of triangle *ABC* is the incenter *I*, as we see in Figure 48-P, so that for angle bisector *CI* we have $\angle ECI = \angle ICB$. Furthermore, $\angle ICB = \angle IDB$ since both angles are measured by arc *BI*. This gives us $\angle ACI = \angle ECI = \angle IDB = \angle IDA$, and because $\angle DAI = \angle IAC$, and triangles *DAI* and *IAC* have a common side *AI*, we obtain $\triangle ADI \cong \triangle AIC$, so that $AC = AD$ and $IC = ID$. Because of the angle equality $\angle IDB = \angle ECI$, we have $\overset{\frown}{BI} = \overset{\frown}{EI}$. Furthermore, since $IC = ID$, we also have $\angle CDI = \angle ICD$, and therefore, $\overset{\frown}{DI} = \overset{\frown}{CI}$. By subtraction, we then obtain $\overset{\frown}{DB} = \overset{\frown}{DI} - \overset{\frown}{BI} = \overset{\frown}{CI} - \overset{\frown}{EI} = \overset{\frown}{CE}$, or $DB = CE$. Since we have already established $AD = AC$, we finally get $AB = AD - BD = AC - EC = AE$.

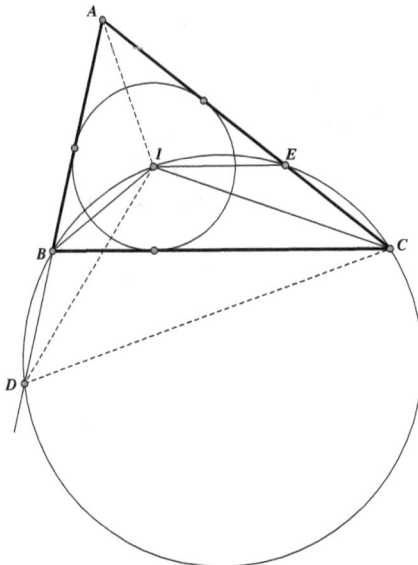

Figure 48-P

## Curiosity 49. Adventures with Circumscribed and Inscribed Circles of a Triangle

In Figure 49-P, we first note that the incenter $I$ of triangle $ABC$ is the common point of the interior angle bisectors. We therefore have $\angle BCI = \frac{1}{2}\angle CBA$ and $\angle ICB = \frac{1}{2}\angle ACB$. Considering triangle $IBC$, we find that

$$\angle BIC = 180° - \angle CIB - \angle ICB = 180° - \frac{1}{2}\angle CBA - \frac{1}{2}\angle ACB.$$

Now, considering triangle $ABC$, we have

$$\angle BIC = 180° - \frac{1}{2}\angle CBA - \frac{1}{2}\angle ACB = 180° - \frac{1}{2}(\angle CBA + \angle ACB).$$

$$= 180° - \frac{1}{2}(180° - \angle BAC) = 90° + \frac{1}{2}\angle BAC.$$

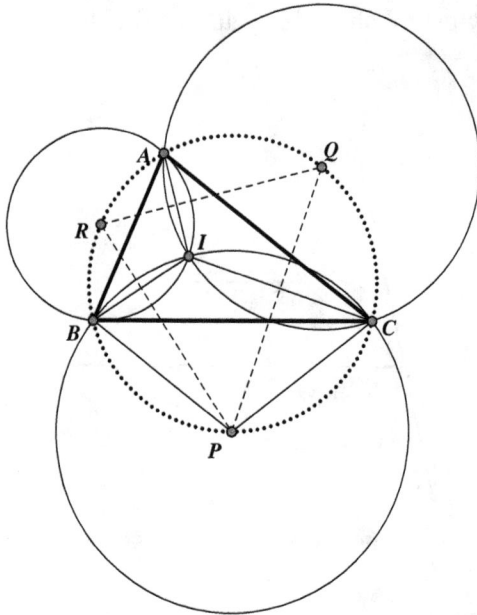

Figure 49-P

Since point $P$ is the circumcenter of triangle $BCI$, we find that $\angle CPB = 2 \cdot (180° - \angle BIC) = 180° - \angle BAC$, and we see that $ABPC$ is a cyclic quadrilateral. Similarly, we can show that quadrilaterals $BCQA$ and $RARB$ are also cyclic. Thus points $P$, $Q$, and $R$ all lie on the circumcircle of triangle $ABC$.

## Curiosity 50. More Adventures with Circumscribed and Inscribed Circles of a Triangle

To begin this proof, we first extend $IP$, $IQ$, and $IR$ to meet the circumscribed circles of triangles $BCI$, $CAI$, and $ABI$ at points $D$, $E$, and $F$, respectively, as shown in Figure 50-P. We then draw the lines $DE$, $EF$, $FD$, $ID$, $IE$, and $IF$. Since $IF$ is a diameter of the circumscribed circle of triangle $ABI$, $\angle IBF = 90°$. Similarly, the circumscribed circle of triangle $BCI$ yields $\angle DBI = 90°$, and $FBD$ is therefore a straight line. In the same way, we can show that $DCE$ and $EAF$ are also straight lines.

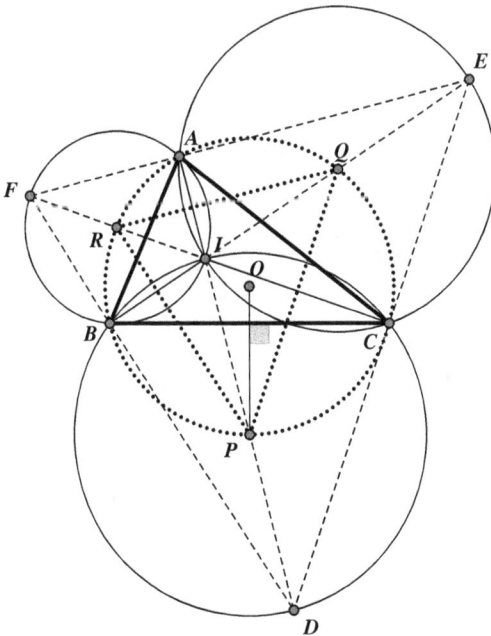

Figure 50-P

From Curiosity 49, we know that point *P* lies on the circumcircle of triangle *ABC*. Since *PC* = *PB*, ∠*BAP* = ∠*PAC* in the circumcircle of *ABC*, and point *P* lies on the bisector of ∠*BAC*. This is also the case for the incenter *I* so that points *A*, *I*, and *P* are collinear, and since points *I*, *P*, and *D* are also collinear, *AIPD* is a straight line. Similarly, *BIQE* and *CIRF* are also straight lines.

We now note that points *P*, *Q*, and *R* are the midpoints of *ID*, *IE*, and *IF*, respectively. The sides of triangle *PQR* are therefore parallel to the sides of triangle *DEF*. Since we have already established that *IB* ⊥ *DF* and *BIQE* is a straight line, this line *BIQF* is an altitude of triangle *DEF*, as well as of triangle *PQR*. Similarly, lines *AIPD* and *CIRF* are also altitudes of triangle *PQR*, and point *I* is therefore the orthocenter of triangle *PQR*.

## Curiosity 51. How Four Centers of a Triangle can Generate an Unusual Cyclic Quadrilateral

We begin in Figure 51-P by drawing the bisector *AID* of ∠*BAC*, which intersects circle *O* at point *P*. We then construct line *OP* to intersect side *BC* at point *Q*. We note that ∠*BAP* = ∠*PAC*, since *P* lies on the angle bisector, and so $\overset{\frown}{BP} = \overset{\frown}{PC}$. Also, since *OP* is then the bisector of *BC*, point *Q* is the midpoint of *BC*. Next, we add the lines *AO*, *BO*, *CO*, *CI*, *CP*, *PH*, *PB*, and *CD*. Considering circle *O*, we have ∠*POC* = 2 · ∠*PAC*, and since *AIP* bisects the 60° angle *BAC*, we have ∠*POC* = ∠*BAC* = 60°. Since *OC* = *OP*, triangle *OPC* is equilateral, and *PC* = *OP*.

In triangle *AIC*, the exterior angle *PIC* is equal to the sum ∠*IAC* + ∠*ACI*. Furthermore, ∠*IAC* = ∠*BAI* and ∠*ACI* = ∠*ICB*. Thus, since ∠*BAI* = ∠*BAP* = ∠*BCP* in circle *O*, ∠*IAC* + ∠*ACI* = ∠*BAI* + ∠*ICB* = ∠*BCP* + ∠*ICB* = ∠*ICP*. This yields ∠*PIC* = ∠*ICP*, making triangle *PCI* isosceles so that *PC* = *PI* = *PO*.

Next, we note that *CI* and *CD* are the internal and external bisectors of angle ∠*ACB*, which makes them perpendicular. As a result, triangle *IDC* is a right triangle, with ∠*ICD* = 90° and *PC* = *PI*, which gives us *PC* = *PI* = *PD*. Since *Q* is the midpoint of *BC* and *OP* ⊥ *BC*, triangles *QBP* and *QPC* are congruent, and *PB* = *PC*.

Finally, we turn our attention to quadrilateral $AHPO$. Both $AH$ and $OP$ are perpendicular to $BC$, so they are parallel. Also, since points $A$ and $P$ lie on circle $O$, $OA = OP$. We now note that $BQ = \frac{1}{2}BC$ and $\angle BOQ = \frac{1}{2}\angle BOC = \angle BAC = 60°$, which gives us $OQ = \frac{BC}{2\sqrt{3}}$. Also, naming the intersection of $AC$ and the altitude $BH$ as point $E$, we see that $\angle HAC = 90° - \angle ACB$ and $\angle EHA = 90° - \angle HAC = \angle ACB$. This gives us $BE = BC \cdot \sin \angle ACB$, $AE = \frac{BE}{\sqrt{3}} = \frac{BC \cdot \sin \angle ACB}{\sqrt{3}}$, and thus, $AH = \frac{AE}{\sin \angle EHA} = \frac{AE}{\sin \angle ACB} = \frac{BC}{\sqrt{3}}$. We see that $AH = 2 \cdot OQ = OP$, and $AHPO$ is therefore a rhombus, yielding $PO = PH$.

In summary, we have shown that $PD = PC = PO = PI = PH = PB$, and point $P$ is the center of a circle containing all points $D$, $C$, $O$, $I$, $H$, and $B$.

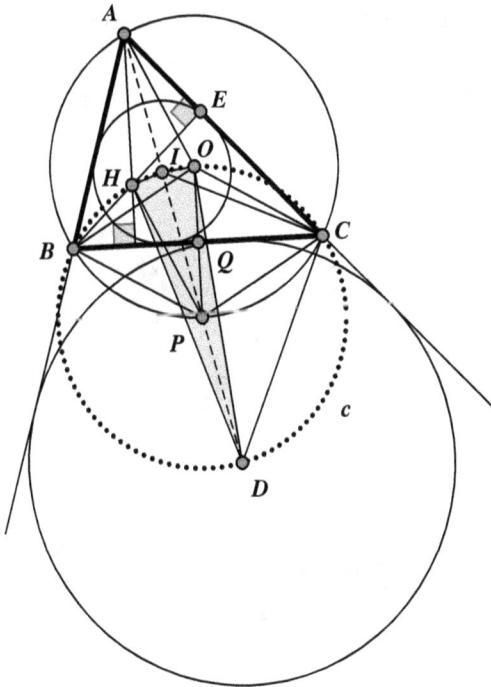

Figure 51-P

## Curiosity 52. The Radii of an Inscribed and Escribed Circle Determine a Rectangle Area

We begin by adding a variety of lines to Figure 52-P. We first recall that, in the proof of Curiosity 50, the angle bisector *AID* of ∠*BAC* intersects the circumcircle of triangle *ABC* in the center *P* of a circle that contains points *B, D, C*, and *I*. We add the line *AIPD* and the circle *P* to the figure.

With the inscribed circle of triangle *ABC* tangent to side *AB* at point *X*, and the escribed circle tangent to side *AB* at point *Y*, we also add the lines *BI, CI, BD*, and *CD*. The extension of *IX* intersects circle *P* at point *F* and the extension of *DY* intersects circle *P* at point *G*. We then also add *FD* and *GI*. The bisectors of the interior and exterior angles of *ABC* at point *B* are *BI* and *BD*, respectively. Similarly, the bisectors of the interior and exterior angles at point *C* are *CI* and *CD*, respectively. *CI* and *CD* are thus perpendicular so that ∠*DBI* = ∠*ICD* = 90°, and *ID* is a diameter of circle *P*. Since points *F* and *G* also lie on this circle, ∠*DFI* = ∠*IGD* = 90°, and *IFDG* is a rectangle.

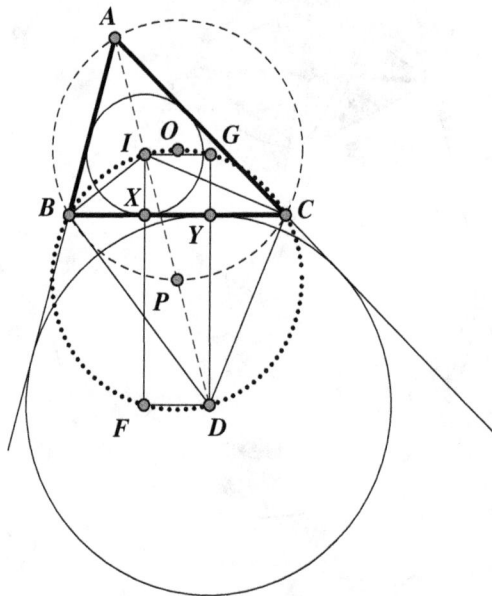

Figure 52-P

In rectangle *IFDG* we have $IX = GY$ and $FX = DY$. Since points $C, G, I, B, F,$ and $D$ all lie on circle $P$, we therefore obtain chord products $CX \cdot BX = IX \cdot XF = IX \cdot DY = GY \cdot DY = CY \cdot BY$. We see that $CX \cdot BX = CY \cdot BY$, with both expressions equal to $IX \cdot DY$, which is the area of a rectangle with sides $IX$ and $DY$.

## Curiosity 53. A Circumscribed Circle and an Inscribed Circle Generate Three Equal-Length Lines

In Figure 53-P, we first note that the incenter $I$ lies on $BP$, which means that $BP$ is the bisector of $\angle CAB$. From this, we get $\overset{\frown}{AP} = \overset{\frown}{CP}$, and therefore, $PA = PC$. Similarly, $\angle BAI = \angle IAC$, since $AI$ bisects $\angle BAC$. The angle $\angle PIA$ is the exterior angle of triangle $ABI$, and $\angle PIA = \angle IBA + \angle BAI$. Furthermore, we have $\angle IAP = \angle CAP + \angle IAC$ and $\angle IBA = \angle PBA = \angle CBP = \angle CAP$, as they are angles inscribed in the same arc $PC$. Thus, we have $\angle PIA = \angle IBA + \angle BAI = \angle CAP + \angle IAC = \angle IAP$, and it follows that triangle $AIP$ is isosceles and $PA = PI$. Thus, we have reached our desired conclusion: $PA = PI = PC$.

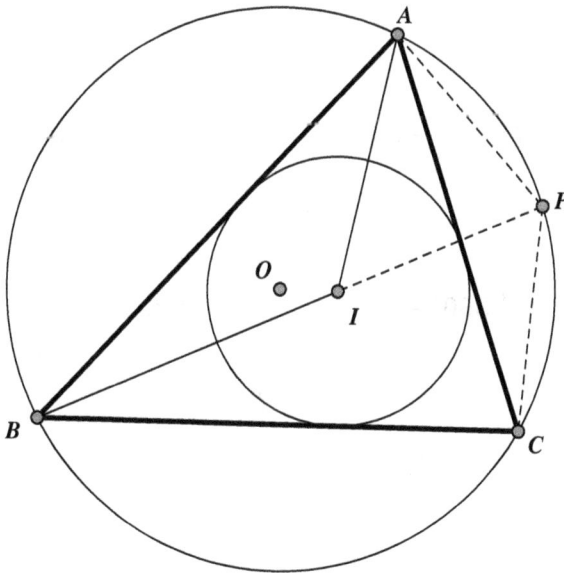

Figure 53-P

## Curiosity 54. Three Circles with a Surprising Common Point

To Figure 54-P, we add lines *OB, OD, OF, HB, HD,* and *HF.* We define point *H* as the one of the intersection points of circles *A* and *C,* and point B as their second point of intersection. Our goal is to show that point *H* will then also lie on the circle with center *E.* We first note in circle *c* that $\angle BAF = 180° - \frac{1}{2}\angle FOB$, and $\angle DCB = 180° - \frac{1}{2}\angle BOD$. This allows us to calculate the following:

In circle *A*, we have

$$\angle FHB = 180° - \frac{1}{2}\cdot\angle BAF = 180° - \frac{1}{2}\left(180° - \frac{1}{2}\angle FOB\right) = 90° + \frac{1}{4}\angle FOB.$$

In circle *C*, we have

$$\angle BHD = 180° - \frac{1}{2}\cdot\angle DCB = 180° - \frac{1}{2}\left(180° - \frac{1}{2}\angle BOD\right) = 90° + \frac{1}{4}\angle BOD.$$

Therefore,

$$\angle DHF = 360° - \angle FHB - \angle BHD$$
$$= 360° - \left(90° + \frac{1}{4}\angle FOB\right) - \left(90° + \frac{1}{4}\angle BOD\right)$$
$$= 180° - \frac{1}{4}\angle FOB - \frac{1}{4}\angle BOD.$$

Since $\angle DOF = 360° - \angle FOB - \angle BOD$, we calculate:

$$\angle FED = 180° - \frac{1}{2}\angle DOF$$
$$= 180° - \frac{1}{2}(360° - \angle FOB - \angle BOD)$$
$$= \frac{1}{2}\angle FOB + \frac{1}{2}\angle BOD.$$

This gives us

$$180° - \frac{1}{2}\angle FED = 180° - \frac{1}{2}\left(\frac{1}{2}\angle FOB + \frac{1}{2}\angle BOD\right)$$

$$= 180° - \frac{1}{4}\angle FOB + \frac{1}{4}\angle BOD = \angle DHF.$$

Since we now know that $\angle DHF = 180° - \frac{1}{2} \cdot \angle FED$, we see that point $H$ lies on circle $E$, and all three circles $A$, $C$, and $E$ have point $H$ in common.

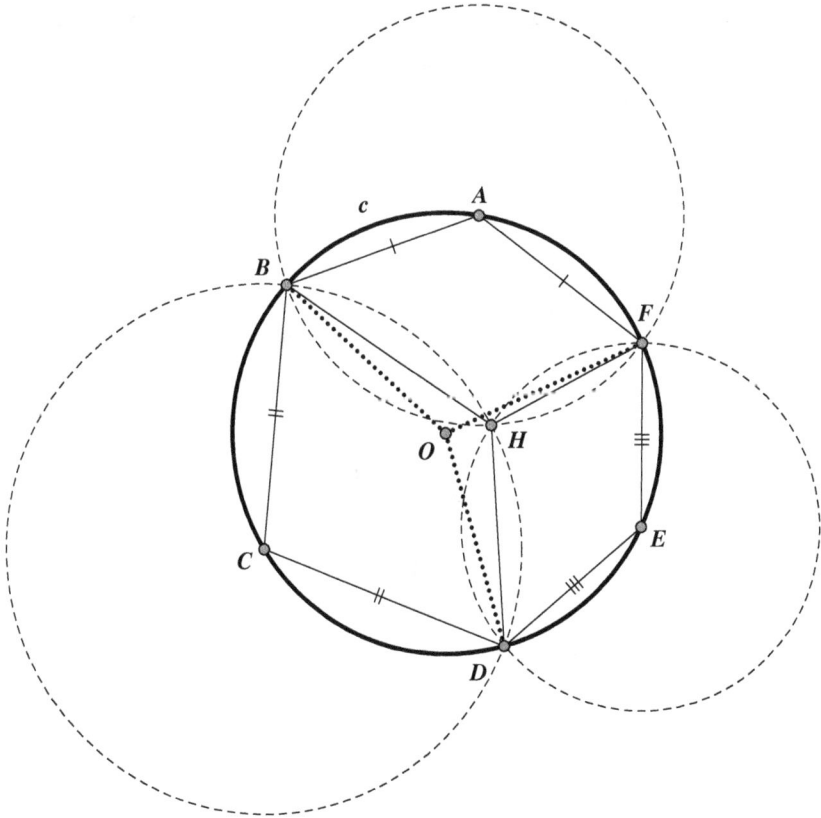

Figure 54-P

## Curiosity 55. The Famous Conway Circle

In Figure 55-P, we have labelled the points at which the incircle is tangent to sides $BC$, $CA$, and $BC$ as $P$, $Q$, and $R$, respectively. Since the tangents to a circle from an exterior point are equal, we label $AQ = AR = x$, $BR = BP = y$, and $CP = CQ = z$. Also, $r$ denotes the radius of the incircle $I$. Considering triangle $IFP$, the radius $IP$ is perpendicular to the side $BC$ at point $P$. Since point $F$ was defined by $BF = CA$, we have $PF = BF + BP = AC + BP = (x+z) + y = x+y+z$, and thus, $IF = \sqrt{IP^2 + PF^2} = \sqrt{r^2 + (x+y+z)^2}$. Similarly, we can also find $IJ = \sqrt{IP^2 + PJ^2} = \sqrt{r^2 + (x+y+z)^2}$, and an analogous argument yields the same values $IE = IH = ID = IG = \sqrt{r^2 + (x+y+z)^2}$ on the extensions of the other two sides of triangle $ABC$. We see that all six points $D$, $E$, $F$, $G$, $H$, and $J$ are at distance $\sqrt{r^2 + (x+y+z)^2}$ from point $I$ and therefore lie on a common circle with center $I$.

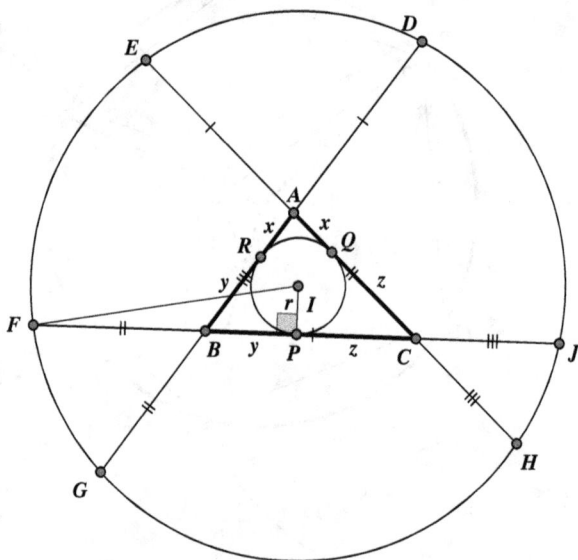

Figure 55-P

## Curiosity 56. A Surprising Collinearity with the Incenter of a Triangle — Sawayama's Lemma

This complex proof requires several steps. As a first step, in Figure 56-P, we extend line $E$ to intersect the circumcircle $O$ of triangle $ABC$ at point $N$. We use a homothety (see Toolbox) with center $D$ to map circle $M$ onto circle $O$ so that the tangent of circle $M$ at point $E$ is mapped onto the tangent of circle $O$ at point $N$. The tangent of circle $O$ at point $N$ is therefore parallel to side $BC$. Point $N$ is then the midpoint of arc $BC$. This yields both $NB = NC$ and $\angle BAN = \angle NAC$, and $AN$ is the bisector of angle $\angle BAC$. Also, since the homothety maps $\triangle DME$ onto $\triangle DON$, we have $\angle EMD = \angle NOD$.

Having collected this preliminary information, we define point $F'$ as the second intersection of line $EI$ with circle $M$. We intend to prove that line $AF'$ is a tangent of circle $M$, and thus $F' = F$.

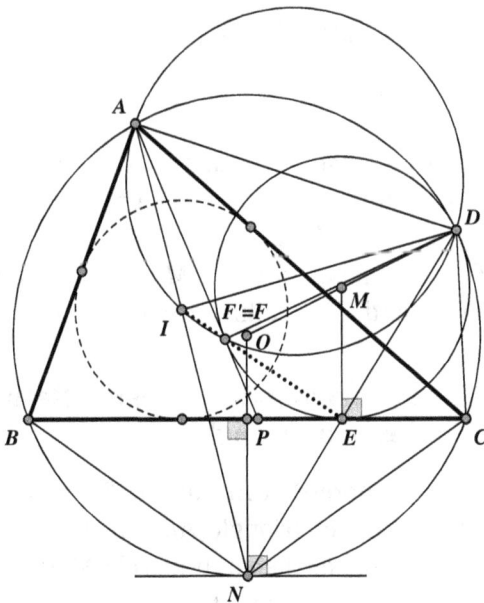

Figure 56-P

We have $\angle EF'D = \frac{1}{2}\angle EMD$ in circle $M$ and $\angle NAD = \frac{1}{2}\angle NOD$ in circle $O$. Since $\angle EMD = \angle NOD$, we get $\angle IAD = \angle NAD = \angle EF'D$. Thus, $\angle DF'I = 180° - \angle EF'D = 180° - \angle IAD$, and quadrilateral $AIF'D$ is cyclic.

Next, we note that triangle $NCB$ is isosceles and quadrilateral $BNCD$ is cyclic, and we obtain $\angle ECN = \angle BCN = \angle NBC = \angle NDC$. This shows us that $\triangle NCE \sim \triangle NCD$ with $\frac{NC}{NE} = \frac{ND}{NC}$.

It will now prove useful to note that $NC = NB = NI$. We can see this by considering triangle $NIB$. In circle $O$, we have $\angle INB = \angle ANB = \angle ACB$ and $\angle NBC = \angle NAC = \frac{1}{2}\angle BAC$. Furthermore, we know that $\angle CBI = \frac{1}{2}\angle CBA$. This means that $\angle NBA = \angle NBC + \angle CBI = \frac{1}{2}\cdot\angle BAC + \frac{1}{2}\cdot\angle CBA$, and in triangle $NIB$ we can calculate

$$\angle BIN = 180° - \angle INB - \angle NBI$$

$$= 180° - \angle ACB - \left(\frac{1}{2}\angle BAC + \frac{1}{2}\angle CBA\right)$$

$$= \frac{1}{2}\angle BAC + \frac{1}{2}\angle CBA$$

$$= \angle NBI.$$

Triangle $NIB$ is thus isosceles with $NI = NB$, and $NC = NB = NI$.

Substituting this into $\frac{NC}{NE} = \frac{ND}{NC}$ gives us $\frac{NI}{NE} = \frac{ND}{NI}$. We see that $\triangle NEI \sim \triangle NDI$, so that $\angle DEF' = \angle DEI = \angle DIA = \angle DF'A$. From this, we see that line $AF'$ is a tangent to circle $M$ so that $F' = F$. In summary, we find that points $E$ and $F$ are collinear with the incenter $I$.

## Curiosity 57. The Surprising Relationship Between the Four Tangent Circles of a Triangle

In Figure 57-P, we first note that $AIE$ and $BIF$ are straight lines that bisect the interior angles of triangle $ABC$ in $A$ and $B$, respectively. With $DN$ being a diameter of the circumcircle $M$ of triangle $DEF$, we draw lines $NF$ and $NE$. We also note that $FAD$, $DBE$, and $ECF$ are also straight lines and the bisectors of the exterior angles of triangle $ABC$

at *A*, *B*, and *C*, respectively. Since the bisectors of the exterior angles are perpendicular to the bisectors of the respective interior angles, we have $AIE \perp DAF$ and $BIF \perp DBE$, and *I* is therefore the ortho-center of triangle *DEF*. Since $\angle NED$ is inscribed in a semicircle, $\angle NED = 90° = \angle FBD$. This means that $NE \parallel FI$, and since we analo-gously have $NF \parallel EI$, it follows that *NFIE* is a parallelogram with $FI = NE$. In right triangle NDE, we obtain $DE^2 + FI^2 = DE^2 + NE^2 = DN^2$. Similarly, we also obtain $DF^2 + EI^2 = DN^2$, and thus, $DE^2 + FI^2 = DF^2 + EI^2 = DN^2$.

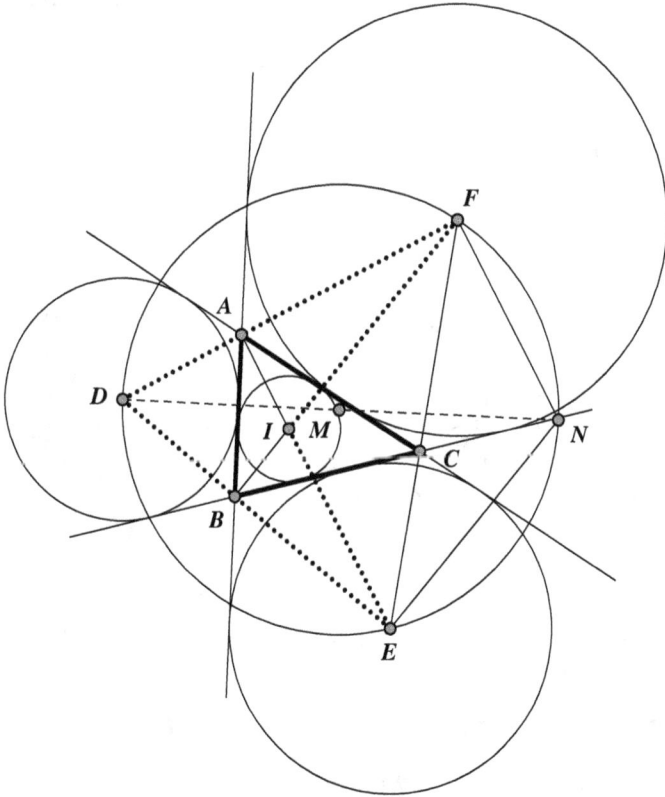

Figure 57-P

## Curiosity 58. Two Tangent Circles Generate an Unexpected Right Angle

As we know, radii of a circle drawn to the points of tangency are always perpendicular to that tangent. In Figure 58-P, this means that we have $OR \perp RS$ and $QS \perp RS$. Since the sum of the angles of a quadrilateral is $360°$, $\angle QOR + \angle SQO = 180°$. Since $\angle QOR = \angle POR = 2\angle PAR$ in circle $O$ and $\angle SQO = \angle SQP = 2\angle SBP$ in circle $Q$, we can then consider triangle $HAB$ to obtain the following:

$$\angle AHB = 180° - (\angle BAH + \angle HBA) = 180° - (\angle PAR + \angle SBP)$$

$$= 180° - \frac{1}{2}(\angle QOR + \angle SQO) = 180° - 90° = 90°,$$

and thus, $\angle AHB = 90°$.

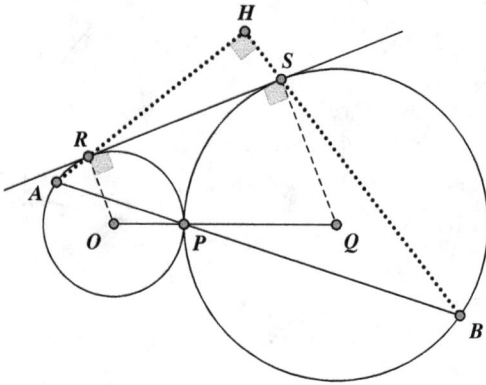

Figure 58-P

## Curiosity 59. The Unexpected Diameter of a Circle

We begin Figure 59-P by drawing the two radii $OA$ and $QB$. The common tangent to the two circles at point $P$ intersects $AB$ at point $C$. Since an angle formed by a chord and a tangent is one-half the intercepted arc, we have $\angle CPA = \frac{1}{2}\angle POA$ and $\angle BPC = \frac{1}{2}\angle BQP$. By addition, we obtain $90° = \angle BPA = \angle CPA + \angle BPC = \frac{1}{2}(\angle POA + \angle BPC)$, and thus,

$\angle POA + BPC = 180°$, or $AO||BQ$. Since $AO = BQ$, we can conclude that $AOQB$ is a parallelogram with $AB = OQ$. Thus, line $AB$ is equal in length to the sum of the radii of the two circles.

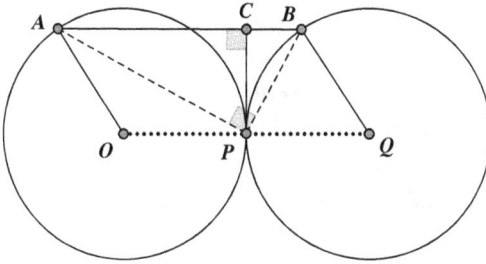

Figure 59-P

## Curiosity 60. Doubling the Distance Between Centers

In Figure 60-P, we draw lines $OM$, $OA$, $QA$, and $QY$, noting that triangles $OAX$ and $QYA$ are isosceles since they each have two sides that are radii in their respective circles. This means that the feet $M$ and $N$ of their altitudes $OM$ and $QN$ are the midpoints of their bases $AX$ and $AY$, respectively. We also note that quadrilateral $OQNM$ is a rectangle, and we therefore have $OQ = MN$. We, therefore, obtain $XY = XA + AY = 2 \cdot MA + 2 \cdot AN = 2 \cdot (MA + AN) = 2 \cdot MN = 2 \cdot OQ$, or $XY = 2 \cdot OQ$.

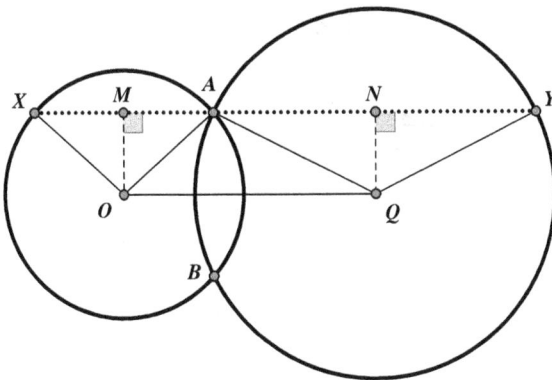

Figure 60-P

### Curiosity 61. Tangent Circles Generating a Circle Tangent to the Line of Centers

We begin by drawing the lines $AP$, $BP$, $AO$, and $BO$ in Figure 61-P. We then draw the common tangent of the two circles at $P$, which intersects $AB$ at point $R$. We note that $RA = RP$ as they are both tangent to circle $O$. Similarly, $RB = RP$ as they are both tangent to circle $Q$. This gives us $RA = RP = RB$, which means that $R$ is the midpoint of $AB$ and, therefore, the center of the circle with diameter $AB$, which also contains point $P$. Now we need to show that this circle is tangent to the line $OPQ$. To that end, we consider triangle $OPA$. Since $OA = OP$, this triangle is isosceles with $\angle OAP = \angle APO$. Since radius $OA$ of circle $O$ is perpendicular to the circle's tangent $AB$, it is a tangent of the circle with diameter $AB$. Therefore, $\angle OAP = \angle ABP$ because both angles have one-half the measure of $\overset{\frown}{AP}$ on circle $R$. This means that we also have $\angle APO = \angle OAP = \angle ABP$, and the circle with diameter $AB$ is thus tangent to line $OPQ$ at point $P$.

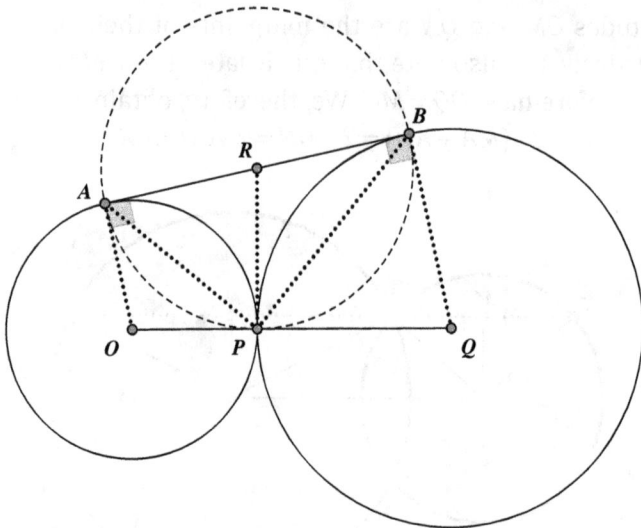

Figure 61-P

## Curiosity 62. An Unexpected Angle Bisector Generated by Externally Tangent Circles

We first draw chord $AB$ and the line $CP$, the common tangent of the two circles at point $P$, as shown in Figure 62-P. Since the angle $\angle XPC$ between a chord of a circle and the tangent in one of its endpoints is equal to half the measure of the intercepted arc $\overset{\frown}{PX}$, we have $\angle XPC = \angle XYP$ in circle $O$ and, analogously, $\angle CPA = \angle PBA$ in circle $Q$. Thus, $\angle XPA = \angle XPC + \angle CPA = \angle XYP + \angle PBA$. Similarly, we also have $\angle PAX = \angle PBA$. Consideration of the angles in $\triangle ABY$ gives us $\angle BAP = 180° - \angle PAX - \angle AYB - \angle YBA = 180° - 2 \cdot \angle PBA - \angle XYP$. Now considering $\triangle ABP$, we obtain

$$\angle APB = 180° - \angle BAP - \angle PBA = 180° - (180° - 2 \cdot \angle PBA - \angle XYP) - \angle PBA = \angle XYP + \angle PBA.$$

This gives us $\angle APB = \angle XYP + \angle PBA = \angle XPA$, which indicates that $PA$ is the bisector of angle $\angle XPB$.

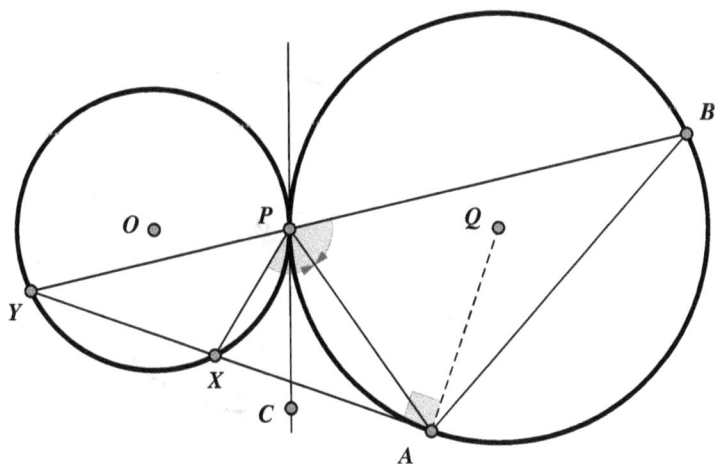

Figure 62-P

## Curiosity 63. Unexpected Parallel Chords in Externally Tangent Circles

We begin this proof by constructing the common tangent $EP$ in Figure 63-P, as well as lines $PX$, $PA$, $OQ$, and $BQ$. In circle $Q$, since $PE$ and $AX$ are tangents and angles $\angle EPA$, $\angle PAX$, and $\angle PBA$ are each measured by one-half arc $AP$, we have $\angle EPA = \angle PAX = \angle PBA$. Similarly, in circle $O$, we have $\angle XYP = \angle XPE$, as these angles are both measured by one-half arc $PX$. We next consider $PX$ in triangle $ABY$. We have established $\angle AYB = \angle XYP = \angle XPE$ and $\angle YBE = \angle PBE = \angle EPA$, and therefore obtain $\angle BAY = 180° - \angle AYB - \angle YBE = 180° - \angle XPE - \angle EPA = 180° - \angle XPA$. Since $\angle QAY = 90°$, we have $\angle BAQ = \angle BAY - \angle QAY = 180° - \angle XPA - 90° = 90° - \angle XPA$. In isosceles triangle $QAB$, we obtain $\angle QAB = 180° - 2 \cdot \angle BAQ = 180° - 2 \cdot (90° - \angle XPA) = 2 \cdot \angle XPA$, which gives us $\angle APB = \frac{1}{2} \cdot \angle AQB = \angle XPA$ in circle $Q$. Since $\angle CPX$ and $\angle APB$ are vertical angles, we obtain $\angle CPY = \angle APB = \angle XPA$, and since quadrilateral $PCYX$ is cyclic, we have opposite angles $\angle YCP = \angle AXP$. Triangles $\triangle APX$ and $\triangle YPC$ are therefore similar, and we have $\angle PYC = \angle PAX$. In summary, we have $\angle BYC = \angle PYC = \angle PAX = \angle PBA = \angle YBA$, and lines $AB$ and $YC$ are parallel.

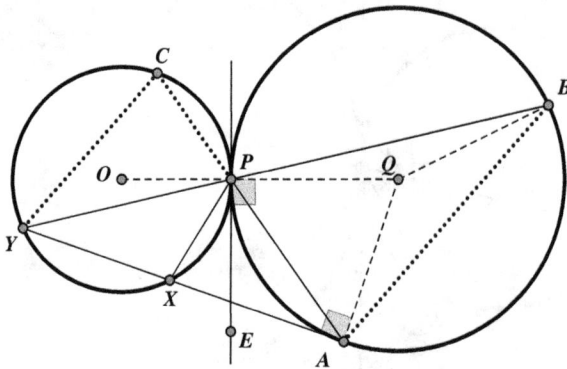

Figure 63-P

## Curiosity 64. Surprising Parallel Tangents of Externally Tangent Circles

In Figure 64-P, we begin by adding the common tangent $XY$ of the two circles at $P$, as well as placing points $C$ and $D$ on the tangent lines at $A$ and $B$, respectively. In circle $O$, we have $\angle CAP = \angle APY$ since the angles are both half the measure of arc $AP$. Similarly, in circle $Q$, we have $\angle DBP = \angle BPX$. The vertical angles $\angle APY$ and $\angle BPX$ are equal, so that we obtain $\angle CAP = \angle DBP$. Since these are the alternate interior angles for the tangent lines at points $A$ and $B$, it follows that they are parallel.

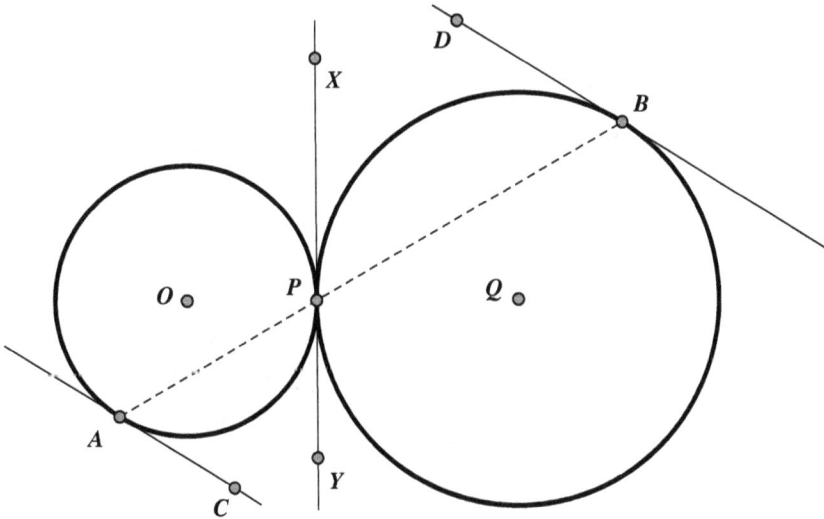

Figure 64-P

## Curiosity 65. Two Externally Tangent Circles Generate a Line Half as Long as Their Line of Centers

We let $M$ denote the midpoint of $OQ$, as shown in Figure 65-P, and let $D$ be the midpoint of $BC$. Since $M$ is the midpoint of $OQ$ and $D$ is the midpoint of $BC$, the right trapezoid $OQCB$ gives us $OB||MD||QC$, and $MD = \frac{1}{2}(OB + QC) = \frac{1}{2}(OA + QA) = \frac{1}{2}OQ$. Since $OB||MD||QC$, we have $MD \perp BC$, and since we also have $AP \perp BC$, it follows that $MD||AP$. We now note that point $D$ is also the intersection of $BC$ with the common tangent of the two circles at point $A$, as we have $DA = DB = DC$. Therefore, we have $DA \perp OQ$ and $MP \perp OQ$, and thus, $DA||MP$. We see that quadrilateral $APMD$ is a parallelogram, and thus, $AP = MD = \frac{1}{2}OQ$, or $OQ = 2AP$.

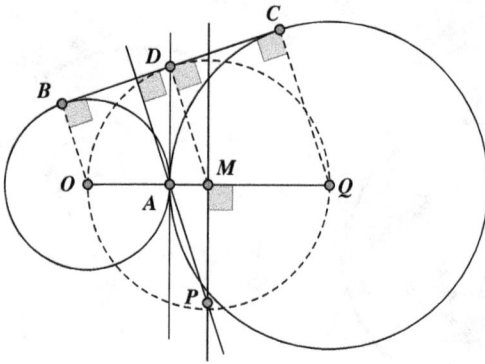

Figure 65-P

## Curiosity 66. A Very Surprising Constant Tangent Length

In this unusual configuration, we will require several additional points and lines, as we see in Figure 66-P. Since point $M$ is the midpoint of arc $AB$, the line joining center $O$ of the given circle with $M$ is perpendicular to chord $AB$. Point $N$ is the intersection of $MO$ with $AB$, and we let $R$ denote the length of the radius $OM$ and $x$ the length of $NM$. Since circle $Q$ is tangent to $AB$ at point $C$, we also have $QC \perp AB$, and can therefore complete the rectangle $OYZM$ with points $Y$ and $Z$

on the extension of $QC$. We let $r$ denote the length of the radius $QC$ of the tangent circle, and since the circles are tangent at point $D$ on $OQ$, we have $OQ = OD + DQ = R + r$. Finally, we also add lines $MQ$ and $QE$, where point $E$ is the point of tangency on circle $Q$ with the tangent from point $M$, and where we have $ME = t$.

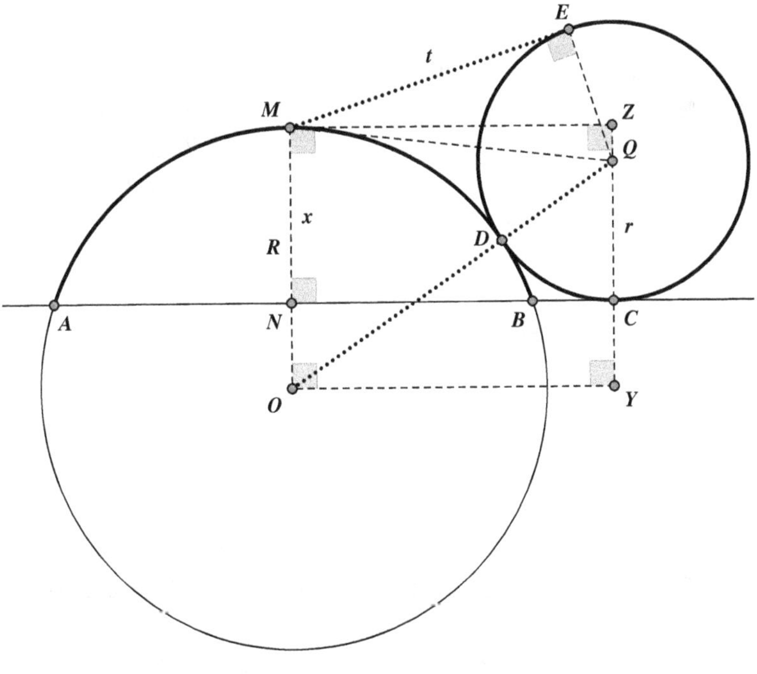

Figure 66-P

Noting that $YQ = YZ - ZC + CQ = OM - MN + CQ = R - x + r$, we can apply the Pythagorean theorem to right triangle $OYQ$ to obtain the following:

$$
\begin{aligned}
OY^2 &= OQ^2 - YQ^2 \\
&= (R+r)^2 - ((R+r) - x)^2 \\
&= (R+r)^2 - (R+r)^2 + 2h(R+r) - x^2 \\
&= 2x(R+r) - x^2.
\end{aligned}
$$

Since $MZ = OY$ in rectangle $OYZM$, we can now apply the Pythagorean theorem to right triangle $ZMQ$, yielding the following:

$$MQ^2 = MZ^2 + ZQ^2$$
$$= 2x(R + r) - x^2 + (x - r)^2$$
$$= 2xR + 2xr - x^2 + x^2 - 2xr + r^2$$
$$= 2xR + r^2.$$

In the final step, we apply the Pythagorean theorem to right triangle $EMQ$, yielding $ME^2 = MQ^2 - QE^2 = 2xR + r^2 - r^2 = 2xR$. We see that the value of $ME^2$, and thus the value of $t = ME$, only depends on the values of $R$ and $x$, but not on the value of $r$. In other words, the value of $t$ only depends on the choice of circle $O$ and secant $AB$, and is independent of the choice of circle $Q$.

## Curiosity 67. Two Circles Creating Parallel Lines Unexpectedly

In Figure 67-P, we have added points $M$ and $N$ on the common tangent of the two circles in $P$ in order to identify the relevant angles. Since $AP$ is a chord of circle $O$ and $PM$ its tangent in $P$, we have $\angle MPA = \angle PDA$. Similarly, since $BP$ is a chord of circle $Q$ and $PN$ its tangent in $P$, we have $\angle NPB = \angle PCB$. We now need only note that vertical angles $\angle MPA$ and $\angle NPB$ are equal to see that $\angle CDA = \angle PDA = \angle MPA = \angle NPB = \angle PCB = \angle DCB$, which shows us that lines $AD$ and $BC$ are parallel.

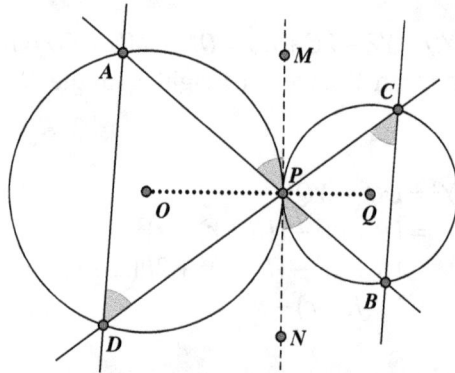

Figure 67-P

## Curiosity 68. Two Circles Create an Angle Bisector

In Figure 68-P, we draw the common tangent *EAF* of the two circles at point *A*, label the intersection of *AB* with circle *Q* point *G*, and draw line *GD*. Since *AB* is a chord of circle *O* and *AE* its tangent in *A*, we have $\angle EAB = \angle ACB$ because both angles are measured by one-half of arc *AB*. Similarly, since *AG* is a chord of circle *Q* and *AE* is tangent to circle *Q* at point *A*, we also have $\angle EAG = \angle ADG$. Since $\angle EAB = \angle EAG$, this gives us $\angle ACB = \angle ADG$ or $\angle ACD = \angle ADG$, since they are each one-half of $\overarc{AG} = \overarc{AB}$.

We now note that *AD* is a chord of circle *Q*, and *DC* its tangent at point *D*. We then have $\angle CDA = \angle DGA$ since both angles are equal to one-half the measure of $\overarc{AD}$. Thus, we have established $\triangle ADC \sim \triangle AGD$ so that the third angles in each of these two triangles must also be equal, namely, $\angle CAD = \angle DAG$. We conclude that *AD* bisects $\angle BAC$.

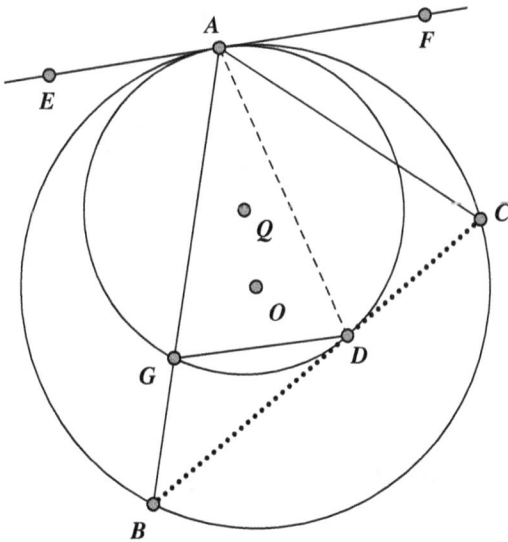

Figure 68-P

### Curiosity 69. Tangent Circles Produce Unusual Line Products

In Figure 69-P, we first draw point $E$ as the second intersection of $AC$ with circle $Q$. Next, we add line segments $GA$ and $DE$, as well as the common tangent $PC$ to the two circles at their common tangency point $C$. Since $PC$ is a common tangent of the two circles, we obtain $\angle CGA = \angle PCA = \angle PCE$, as these angles are all one-half the measure of arc $AC$ of circle $O$. We also have $\angle PCE = \angle CDE$ because they measured by one-half arc $EC$ of circle $Q$. Therefore, since $\angle CGA = \angle CDE$, we have $GA$ and $DE$ parallel, and $\angle GAC = \angle DEC$. Since $ADB$ is a tangent of circle $Q$, we also have $\angle DEC = \angle BDC$, and thus, $\angle GAC = \angle BDC$. Also, since both angles are measured by arc $CA$ in circle $O$, we also have $\angle CBD = \angle CBA = \angle CGA$. Triangles $AGC$ and $DBC$ are therefore similar, and we have $\frac{AC}{CD} = \frac{CG}{BC}$, or $AC \cdot BC = CD \cdot CG$.

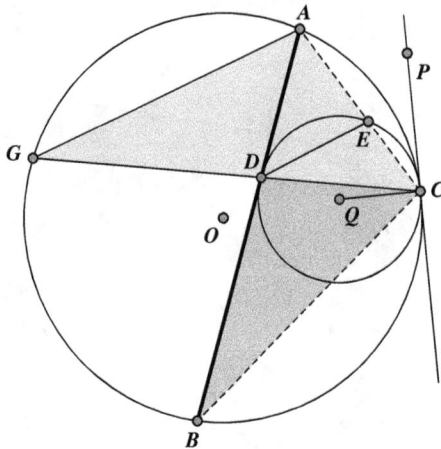

Figure 69-P

### Curiosity 70. Equal Angles Appear in Internally Tangent Circles

We add the common tangent $XY$ of the two circles at point $P$, as shown in Figure 70-P. We now note that $\angle CPA = \angle CPX - \angle APX$. We have $\angle CPX = \angle CDP$ since both angles are measured by half the measure of arc $PC$. Analogously, $\angle APX = \angle ABP$ in circle $O$. Therefore, $\angle CPA = \angle CPX - \angle APX = \angle CDP - \angle ABP$. Now considering triangle $PBD$, this gives us $\angle CPA = \angle CDP - \angle ABP = (180° - \angle PDB) - \angle DBP = \angle BPD$, and we have $\angle CPA = \angle BPD$.

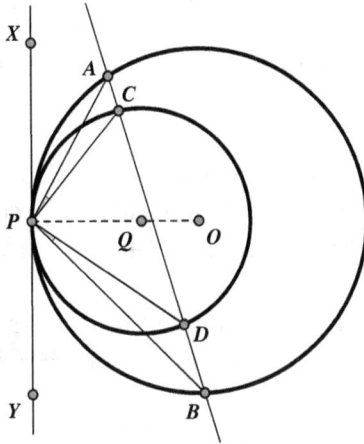

Figure 70-P

## Curiosity 71. Incredible Tangent Lengths in Tangent Circles

In Figure 71-P, we first extend $PO$ to meet circle $O$ a second time at $P'$. We define the radius of circle $O$ as $R$, the radius of circle $Q$ as $r$, and $\angle P'OB = x$, noting that $x < 60°$. Before we start our calculations, we also note that $OQ = R - r$, $\angle QOA = 180° - \angle AOP' = 180° - (120° - x) = 60° + x$, $\angle BOQ = 180° - \angle P'OB = 180° - x$, and $\angle COQ = 180° - \angle P'OB - \angle BOC = 180° - x - 120° = 60° - x$.

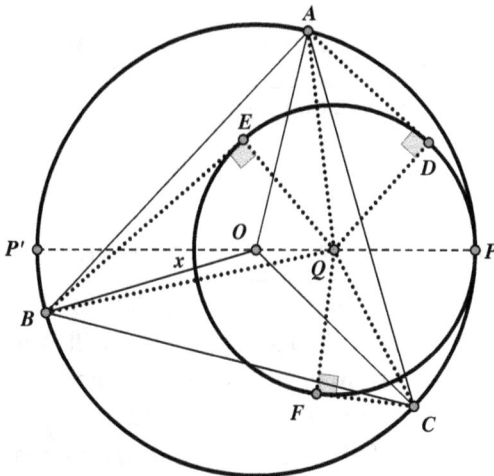

Figure 71-P

We apply the law of cosines (see Toolbox) to triangle $OBQ$ to get $BQ^2 = OB^2 + OQ^2 - 2 \cdot OB \cdot OQ \cdot \cos \angle BOQ$, or in terms of length measures

$$BQ^2 = R^2 + (R-r)^2 - 2R(R-r) \cdot \cos(180° - x)$$
$$= R^2 + R^2 - 2Rr + r^2 + 2R(R-r) \cdot \cos x$$
$$= 2R(R-r)(1 + \cos x) + r^2.$$

The Pythagorean theorem applied to right triangle $BQE$ gives us $BE^2 = BQ^2 - QE^2 = (2R(R-r)(1 + \cos x) + r^2) - r^2 = 2R(R-r)(1 + \cos x)$.

Similarly, again applying the law of cosines to triangle $OQA$ gives us $AQ^2 = OA^2 + OQ^2 - 2 \cdot OA \cdot OQ \cdot \cos \angle QOA$, or $AQ^2 = R^2 + (R-r)^2 - 2R(R-r) \cdot \cos(60° + x) = 2R(R-r)(1 - \cos(60° + x)) + r^2$.

Applying the Pythagorean theorem to right triangle $AQD$ gives us $AD^2 = AQ^2 - QD^2 = (2R(R-r)(1 - \cos(60° + x)) + r^2) - r^2 = 2R(R-r)(1 - \cos(60° + x))$.

Furthermore, applying the law of cosines to triangle $OCQ$ gives us $CQ^2 = OC^2 + OQ^2 - 2 \cdot OC \cdot OQ \cdot \cos \angle COQ$, or $CQ^2 = R^2 + (R-r)^2 - 2R(R-r) \cdot \cos(60° - x) = 2R(R-r)(1 - \cos(60° - x)) + r^2$.

Applying Pythagorean theorem in right triangle $CQF$ gives us $CF^2 = CQ^2 - QF^2 = (2R(R-r)(1 - \cos(60° - x)) + r^2) - r^2 = 2R(R-r)(1 - \cos(60° - x))$.

We need to show $BE = AD + CF$, and this can be written as

$$\sqrt{2R(R-r)(1 + \cos x)} =$$
$$\sqrt{2R(R-r)(1 - \cos(60° + x))} +$$
$$\sqrt{2R(R-r)(1 - \cos(60° - x))}.$$

Dividing by the factor $\sqrt{2R(R-r)}$, this is equivalent to $\sqrt{1 + \cos x} = \sqrt{1 - \cos(60° + x)} + \sqrt{1 - \cos(60° - x)}$. In order to show that this is true for any value of $x$ with $0 \leq x \leq 60°$, we will require some elementary rules of addition for the cosine function (see Toolbox): $\cos(a + b) = \cos a \cdot \cos b - \sin a \cdot \sin b$ and $\cos(a - b) = \cos a \cdot \cos b + \sin a \cdot \sin b$. Substituting 60° for $a$ and $x$ for $b$, these rules give us $\cos(60°+x)=\cos 60°\cdot\cos x-\sin 60°\cdot\sin x=\frac{1}{2}\cdot\cos x-\frac{\sqrt{3}}{2}\cdot\sin x$

and $\cos(60°-x)=\cos 60°\cdot\cos x+\sin 60°\cdot\sin x=\frac{1}{2}\cdot\cos x+\frac{\sqrt{3}}{2}\cdot\sin x$.

The equality $\sqrt{1+\cos x}=\sqrt{1-\cos(60°+x)}+\sqrt{1-\cos(60°-x)}$

that we wish to prove can also be written as

$\sqrt{1+\cos x}=\sqrt{1-\left(\frac{1}{2}\cdot\cos x-\frac{\sqrt{3}}{2}\cdot\sin x\right)}+\sqrt{1-\left(\frac{1}{2}\cdot\cos x+\frac{\sqrt{3}}{2}\cdot\sin x\right)}$    or

$\sqrt{1+\cos x}=\sqrt{\left(1-\frac{1}{2}\cdot\cos x\right)+\frac{\sqrt{3}}{2}\cdot\sin x}+\sqrt{\left(1-\frac{1}{2}\cdot\cos x\right)-\frac{\sqrt{3}}{2}\cdot\sin x}$.    Since

the expressions on both sides of the equation are positive, we can square both sides to obtain the equivalent equation

$$1+\cos x$$

$$=\left(1-\frac{1}{2}\cdot\cos x\right)+\frac{\sqrt{3}}{2}\cdot\sin x+\left(1-\frac{1}{2}\cdot\cos x\right)-\frac{\sqrt{3}}{2}\cdot\sin x+$$

$$2\cdot\sqrt{\left(1-\frac{1}{2}\cdot\cos x\right)^2-\left(\frac{\sqrt{3}}{2}\cdot\sin x\right)^2}$$

$$=2-\cos x+2\cdot\sqrt{1-\cos x+\frac{1}{4}\cdot\cos^2 x-\frac{3}{4}\cdot\sin^2 x}$$

$$=2-\cos x+2\cdot\sqrt{1-\cos x+\frac{1}{4}\cdot\cos^2 x-\frac{3}{4}\cdot\left(1-\cos^2 x\right)}$$

$$=2-\cos x+2\cdot\sqrt{\frac{1}{4}-\cos x+\cos^2 x}$$

$$=2-\cos x+2\cdot\left(\cos x-\frac{1}{2}\right)$$

$$=1+\cos x.$$

This is true for values of $x$ with $0\le x\le 60°$, and we have therefore shown that $\sqrt{1+\cos x}=\sqrt{1-\cos(60°+x)}+\sqrt{1-\cos(60°-x)}$ is true, which proves that $BE=AD+CF$.

## Curiosity 72. Symmetric Equal Angles

Since the center $O$ lies on $ER$, as shown in Figure 72-P, we have $\angle EAR = 90°$. Similarly, since the center $Q$ lies on $ER$, we also have $\angle FAS = 90°$. This gives us $\angle EAF = \angle EAR - \angle FAR = 90° - \angle FAR = \angle FAS - \angle FAR = \angle RAS$, and we then have $\angle EAF = \angle RAS$.

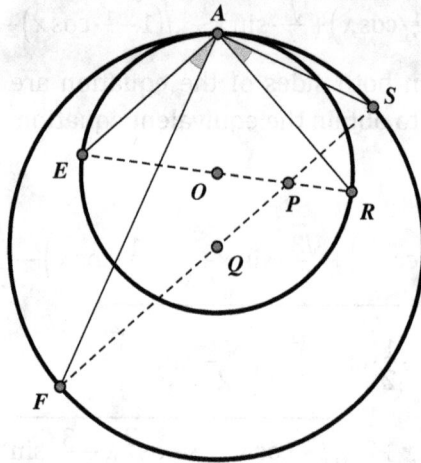

Figure 72-P

## Curiosity 73. An Amazing Rolling Result

To the left side of Figure 73-P we add the radii $QD$ and $QE$ of circle $Q$, so that $\overset{\frown}{AB} = 2\pi OA \frac{\angle AOB}{360°}$ and $\overset{\frown}{AD} = 2\pi QA \frac{\angle AQD}{360°}$. Since $QA = \frac{1}{2}OA$ and $\angle AQD = 2\angle AOD = 2\angle AOB$ in circle $Q$, we obtain $\overset{\frown}{AD} = 2\pi QA \frac{\angle AQD}{360°} = 2\pi \frac{1}{2}OA \frac{2\angle AOB}{360°} = 2\pi OA \frac{\angle AOB}{360°} = \overset{\frown}{AB}$, which gives us $\overset{\frown}{AD} = \overset{\frown}{AB}$. We can show $\overset{\frown}{AE} = \overset{\frown}{AC}$ in an analogous way, and thus, $\overset{\frown}{DE} = \overset{\frown}{AE} - \overset{\frown}{AD} = \overset{\frown}{AC} - \overset{\frown}{AB} = \overset{\frown}{BC}$, or $\overset{\frown}{DE} = \overset{\frown}{BC}$.

We now consider the right side of Figure 73-P. When circle $Q$ rolls up in such a way that points $B$ and $D$ coincide, the tangents of the two circles are such that points coincide with $B = D$, and both radii $OB$ and $QD$ are perpendicular to this common tangent. With point $Y$ on the tangent at point $D$ of circle $Q$, we also note that $\angle YDX$ is the same in both parts of the figure, and we name this angle $x$.

In right triangle $DXA$ on the left side figure, we have $XD = OA \cdot \cos\angle AXD$. Since triangle $QDO$ is isosceles with $QD = QO$, we have $x = \angle YDO = \angle YDQ - \angle ODQ = 90° - \angle QOD = 90° - \angle AXD$; therefore, $XD = OA \cdot \cos(90° - x)$. Now, in the right-side figure, we have triangle $XOD$ with $\angle XDO = 90° - \angle YDX = 90° - x$, $OD = OA$, and $XD = OA \cdot \cos(90° - x)$. This means that $DXOD$ is a right triangle. Since we have already established $\angle AOB = \angle XDO = 90° - x$, we have $XB \| OA$, and point $X$ therefore lies on the diameter on circle $O$ perpendicular to $OA$.

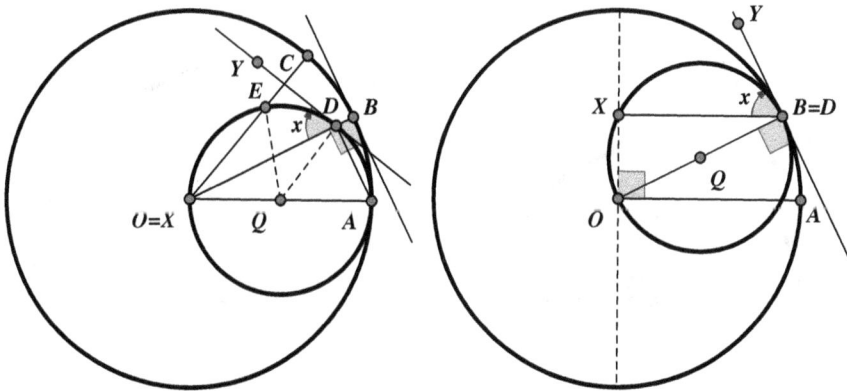

Figure 73-P

## Curiosity 74. Semicircles — The Arbelos

In Figure 74-P, points $A$ and $B$ are the endpoints of the large arc, and point $C$ is the common interior endpoint of the two smaller arcs. Point $O$ is the midpoint of $AB$, point $P$ is the midpoint of $AC$, and point $Q$ is the midpoint of $BC$. The radius of the semicircle with center $P$ is $r_1$ and the radius of the semicircle with center $Q$ is $r_2$. Thus, we have $AB = AC + CB = 2 \cdot r_1 + 2 \cdot r_2 = 2 \cdot (r_1 + r_2)$, and the radius of circle $O$ is $r_1 + r_2$. We know that the arc length of a semicircle is equal to $\pi$ times the circle radius. Therefore, the length of arc $AC$ is equal to $\pi r_1$, the length of arc $BC$ is equal to $\pi r_2$, and the length of arc $AB$ is equal to $\pi(r_1 + r_2)$. Since $\pi \cdot (r_1 + r_2) = \pi \cdot r_1 + p \cdot r_2$, we obtain $\widehat{AB} = \widehat{AC} + \widehat{BC}$.

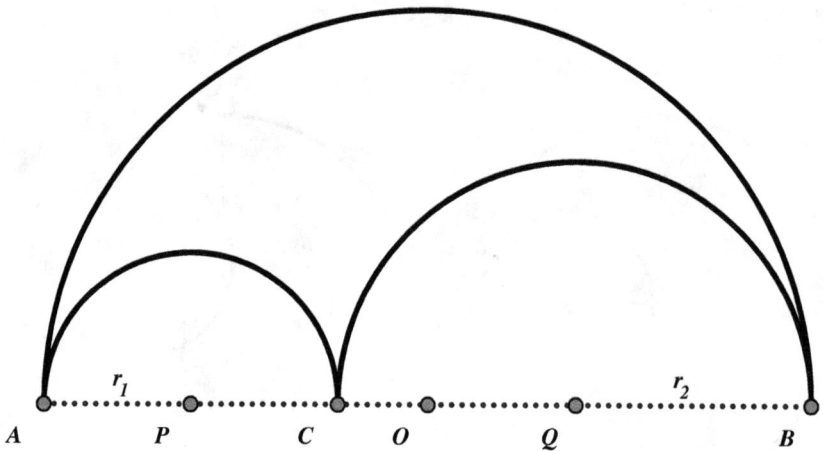

Figure 74-P

## Curiosity 75. An Arbelos Bisection

In Figure 75-P, point $S$ is the intersection of the tangents $FG$ and $HC$ of the two smaller semicircles. Since tangents to a circle from any external point are always of equal length, we have $SC = SF$ for circle $P$ and $SC = SG$ for circle $Q$, so that $SC = SF = SG$. Next, we consider triangle $AHB$. Since point $H$ lies on the circle with the diameter $AC$, it follows that $\triangle AHB$ is a right triangle with the altitude $HC$, which gives us $HC^2 = AC \cdot BC = 2r_1 \cdot 2r_2 = 4r_1r_2$. Now consider quadrilateral $FPJG$ with point $J$ on radius $QG$ in such a way so that $PJ \perp PF$. Since the tangent $FG$ is perpendicular to the radii $PF$ and $GJ$, it follows that $FPJG$ is a rectangle; therefore, $FG = PJ$. We also have $JQ = r_2 - r_1$ and $PQ = r_2 + r_1$. If we apply the Pythagorean theorem to the right triangle $PQJ$, we get $PJ^2 = (r_2 + r_1)^2 - (r_2 - r_1)^2 = 4r_1r_2$, and this gives us $FG^2 = 4r_1r_2$. Thus, $HC = FG$, and together with $SC = SF = SG$, we have $SC = SH$.

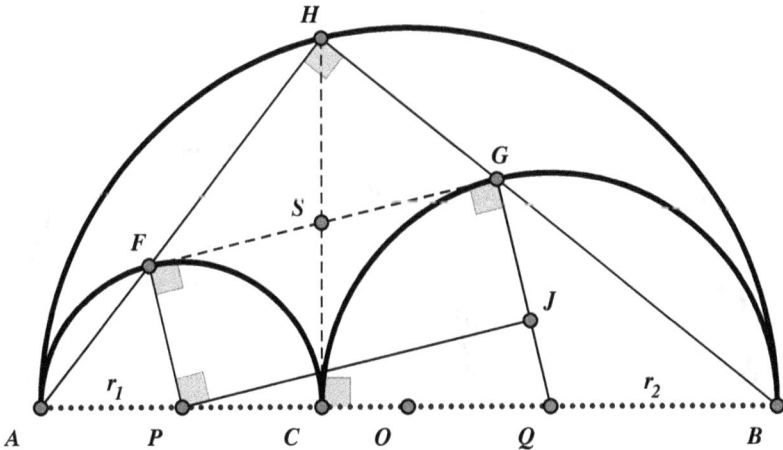

Figure 75-P

## Curiosity 76. More on the Common Arbelos Bisection Point

In the proof of Curiosity 75, we showed that $SC = SF = SG = SH$, so that point $S$ is the center of a circle containing the four points $F$, $C$, $G$, and $H$, as shown in Figure 76-P.

The area of the arbelos can be found by taking the area of the largest semicircle and subtracting the areas of the two smaller semicircles. Since we have $AB = 2r_1 + 2r_2$, the radius of the large semicircle is equal to $r_1 + r_2$, and the area of the arbelos is then equal to

$\frac{\pi(r_1+r_2)^2}{2} - \left(\frac{\pi r_1^2}{2} + \frac{\pi r_2^2}{2}\right) = \frac{\pi}{2} \cdot \left((r_1 + r_2)^2 - r_1^2 - r_2^2\right) = \frac{\pi}{2} \cdot 2r_1 r_2 = \pi \cdot r_1 r_2$. As we saw

in the proof of Curiosity 75, the diameter $FG$ of the circle with center $S$ is equal to $2\sqrt{r_1 r_2}$, and its area is, therefore, equal to $\pi r_1 r_2$, which is the same as the area of the arbelos.

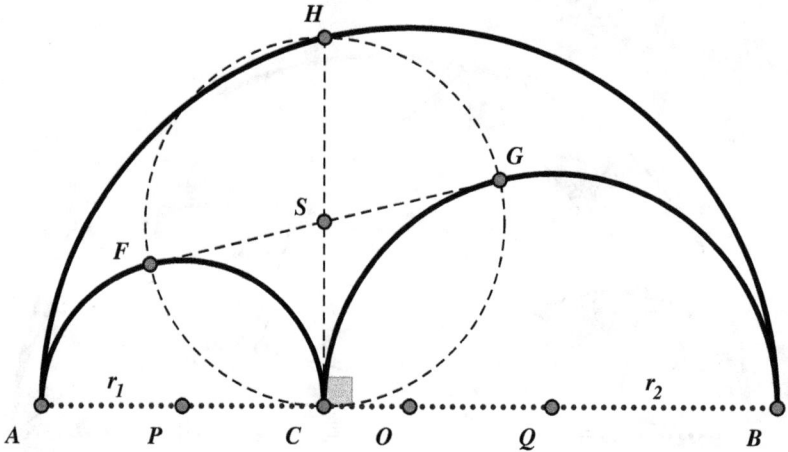

Figure 76-P

## Curiosity 77. Yet More on the Arbelos

In Figure 77-P, we label the intersection of $AH$ with circle $P$ as point $F'$ and label the intersection of $BH$ with circle $Q$ as point $G'$. We then draw lines $F'C$ and $G'C$. Since angles inscribed in a semicircle are right angles, we have $\angle AF'C = \angle AHB = \angle CG'B = 90°$. Quadrilateral $HF'CG'$ is therefore a rectangle, and the diagonal $CH$ is a diameter of the circumcircle of this rectangle. As we have already established in the proof of Curiosity 75, this circle contains points $F$ and $G$, and since these must be the intersections of the circumcircle of $HF'CG'$ with circles $P$ and $Q$, respectively, we obtain $F = F'$ and $G = G'$. This establishes the collinearities $AFH$ and $BGH$.

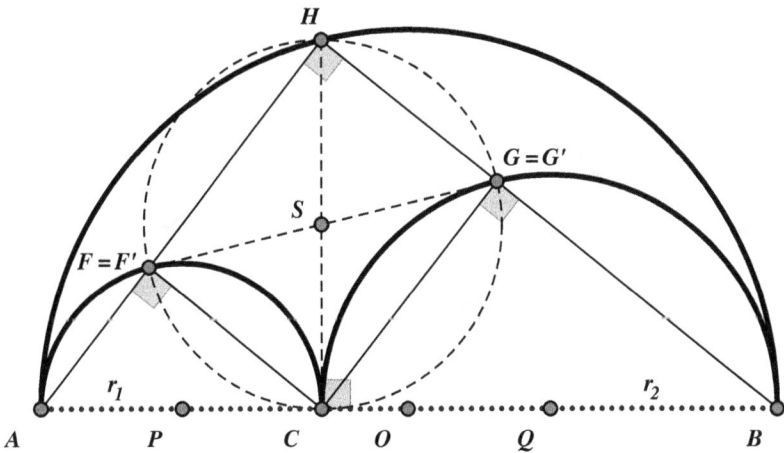

Figure 77-P

## Curiosity 78. Another Arbelos Surprise

We begin by adding lines $AS$, $AR$, and $CR$, as shown in Figure 78-P. Since $\angle ASC$ is inscribed in a semicircle, it is a right angle, and since point $S$ is the midpoint of its semicircular arc, we have $\angle CAS = 45°$. Similarly, since point $R$ is the midpoint of its semicircular arc, we also have $\angle RAC = 45°$, thereby, making angle $\angle SAR = \angle SAC + \angle CAR$ a right angle. We have $SC \| AR$, so that the triangles $SRC$ and $SAC$ share a common base $SC$ and have the same altitude $SA$, which means that they have equal areas. We can now easily calculate area$[SRC] =$ area$[SAC] = \frac{1}{2} \cdot AC \cdot PS = \frac{1}{2} \cdot 2r_1 \cdot r_1 = r_1^2$. In an analogous fashion, it can also be shown that area$[TCR] =$ area$[TCB] = r_2^2$, and consequently, the area of the shaded quadrilateral is equal to area$[SRTC] =$ area$[SRC] +$ area$[TCR] = r_1^2 + r_2^2$.

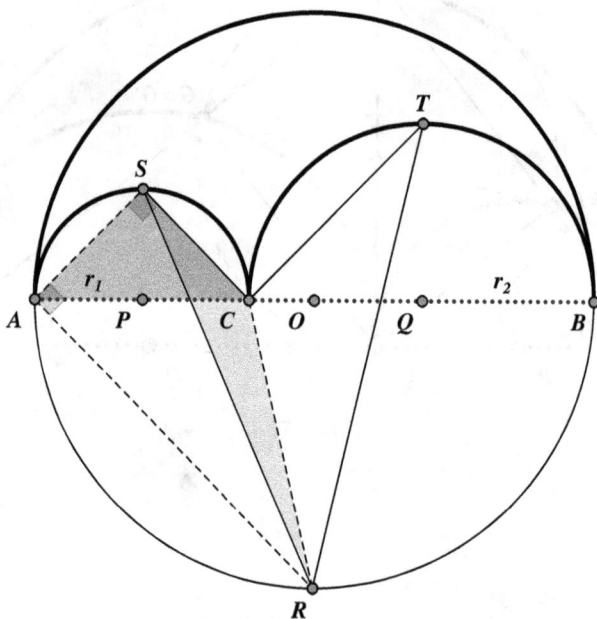

Figure 78-P

## Curiosity 79. Some Further Entertainment with the Arbelos

Points $P$ and $Q$ and radii $r_1$ and $r_2$ are labelled in Figure 79-P as the centers and radii of the two smaller semicircles. The area of the lower semicircle is $\frac{\pi}{2}r_1^2$. The area of the upper portion equals the area of the large semicircle minus the area of the smaller semicircle, or $\frac{\pi}{2}\cdot(r_1+r_2)^2 - \frac{\pi}{2}\cdot r_2^2 = \frac{\pi}{2}\cdot(r_1^2+2r_1r_2+r_2^2) - \frac{\pi}{2}\cdot r_2^2 = \frac{\pi}{2}\cdot(r_1^2+2r_1r_2)$. The area of the entire shape (with the bold border) is thus $\frac{\pi}{2}\cdot r_1^2 + \frac{\pi}{2}\cdot(r_1^2+2r_1r_2) = \frac{\pi}{2}\cdot(2r_1^2+2r_1r_2) = \pi(r_1^2+r_1r_2)$. In order to find the area of the circle $R$, we apply the Pythagorean theorem to right triangle $TAQ$ to obtain $(2RA)^2 + r_2^2 = (2r_1+r_2)^2 = 4r_1^2 + 4r_1r_2 + r_2^2$. This can be written as $4RA^2 = 4r_1^2 + 4r_1r_2$, or $RA^2 = r_1^2 + r_1r_2$. Thus, the area of circle $R$ is equal to $\pi RA^2 = \pi(r_1^2+r_1r_2)$, which is also the area of the shape with the bold border.

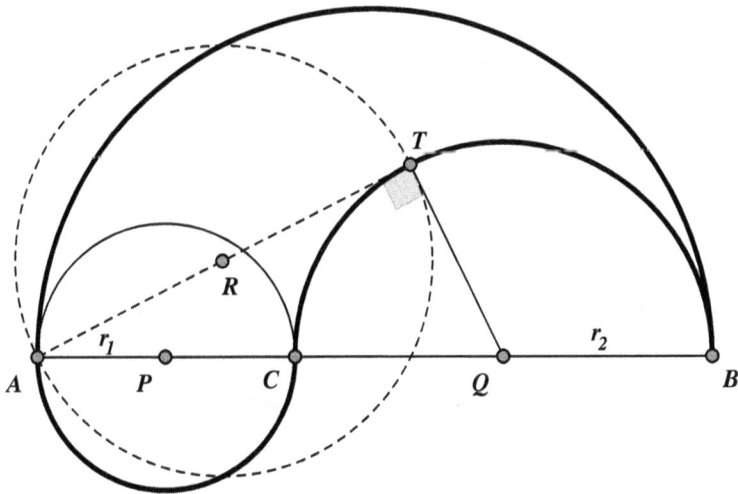

Figure 79-P

## Curiosity 80. A Surprising Area Equality Created by Semicircles

In Figure 80-P, we have labeled areas $m$ and $n$, as well as $t_1 = \text{area}[DCA]$, and $t_2 = \text{area}[DBC]$. We notice that $t_1 + t_2 = \text{area}[ABC]$, $m + y + t_1$ comprises the area of semicircle $ADC$, and $n + x + t_2$ comprises the area of semicircle $CDB$. By the Pythagorean theorem, we know that the sum of the areas of the semicircles on the legs of a right triangle is equal to the area of the semicircle on hypotenuse. Since the area of semicircle $AB$ is $m + n + s$, we have $(m + y + t_1) + (n + x + t_2) = m + n + s$, which simplifies to $s - x - y = t_1 + t_2$ or $s - x - y = \text{area}[ABC]$.

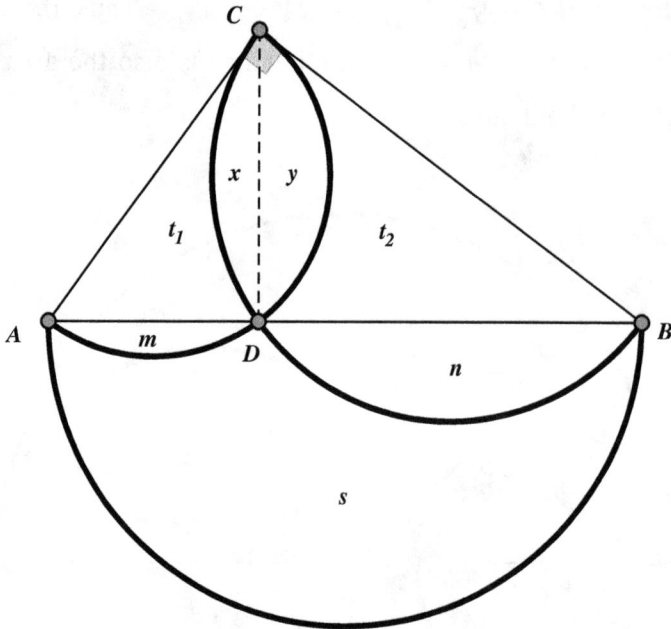

Figure 80-P

## Curiosity 81. Another Surprising Area Equality Created by Semicircles

In Figure 81-P, we have labeled areas $m$, $n$, $x$, and $y$, as well as $t_1 = \text{area}[DCA]$ and $t_2 = \text{area}[DBC]$. As in the proof of Curiosity 80, we again use the fact that the sum of the areas of the semicircles on the legs of a right triangle is equal to the area of the semicircle on the hypotenuse. For triangle $DCA$, this gives us $(m+p) + (y+j) = t_1 + m + y$, which simplifies to $p + j = t_1$. For triangle $DBC$, this gives us $(n+q) + (x+k) = t_2 + n + x$, which simplifies to $q + k = t_2$. We add these two equalities to get $j + k + p + q = t_1 + t_2$, and since $t_1 + t_2 = \text{area}[ABC]$, we obtain $j + k + p + q = \text{area}[ABC]$.

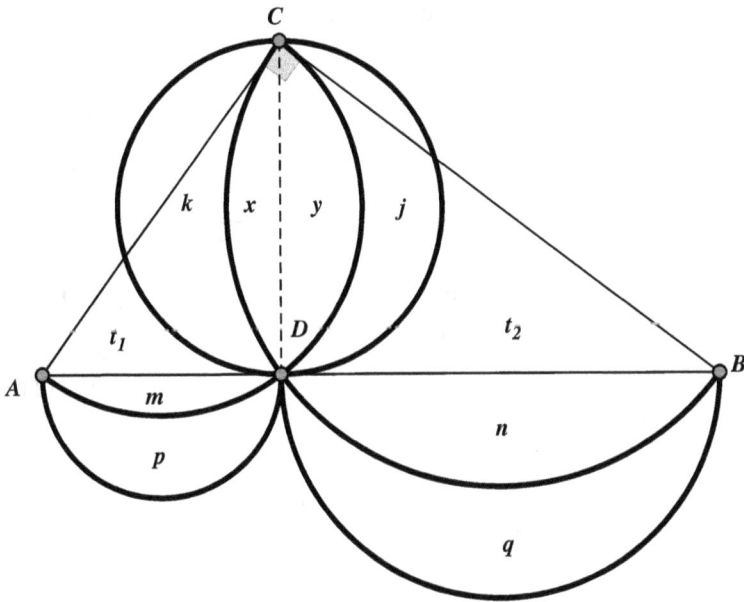

Figure 81-P

## Curiosity 82. An Amazing Curve Division into Two Equal Parts

In Figure 82-P, we label the midpoint $N$ of $MB$ and draw the line $NP$. We let $x$ denote the radius of the small semicircles, and since their diameters $AM$ and $BM$ are radii of the large semicircle, the radius of the large semicircle is $2x$. The vertical angles $\angle AMQ$ and $\angle BMP$ are equal, and we get $\angle BNP = 2\angle BMP = 2\angle AMQ$ in semicircle $N$. We can now express the lengths of the various arcs in terms of $x$ and $\angle AMQ$. We have $\overarc{AM} = 2x\pi$, $\overarc{AQ} = AB \cdot \pi \frac{\angle AMQ}{180°} = \frac{1}{180°} 4x\pi \angle AMQ$, $\overarc{QB} = 4x\pi - \overarc{AQ} = 4x\pi - \frac{1}{180°} 4x\pi \angle AMQ = 4x\pi \left(1 - \frac{1}{180°} \angle AMQ\right)$, $\overarc{BP} = MB \cdot \pi \frac{\angle BNP}{180°} = \frac{1}{180°} 2x\pi \left(2\angle AMQ\right) = \frac{1}{180°} 4x\pi \angle AMQ$, and lastly $\overarc{PM} = 2x\pi - \overarc{BP} = 2x\pi - \frac{1}{180°} 4x\pi \angle AMQ = 2\pi \left(1 - \frac{2}{180°} \angle AMQ\right)$.

The total length of the curve from point $Q$ to point $P$ through point $B$ is equal to

$$\overarc{QB} + \overarc{BP} = 4x\pi \left(1 - \frac{1}{180°} \angle AMQ\right) + \overarc{QB}$$

$$= 4x\pi - \overarc{AQ} = 4x\pi - \frac{1}{180°} 4x\pi \angle AMQ + \frac{1}{180°} 4x\pi \angle AMQ = 4x\pi,$$

and the total length of the curve from point $P$ to point $Q$ and through points $M$ and $A$ is equal to $\overarc{PM} + \overarc{MA} + \overarc{AQ} = 2x\pi \left(1 - \frac{2}{180°} \angle AMQ\right) + 2x\pi + \frac{1}{180°} 4x\pi \angle AMQ = 4x\pi$. Therefore, the lengths of both parts of the curve are equal to $4x\pi$.

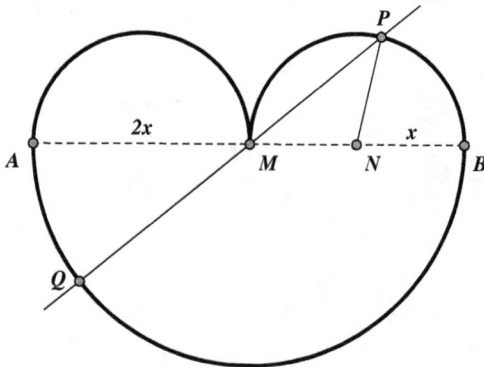

Figure 82-P

## Curiosity 83. An Unexpected Bisection

In Figure 83-P, we draw lines $PA$, $PB$, $QB$, and $CB$. Since arcs $\overset{\frown}{BR}, \overset{\frown}{RQ}$, and $\overset{\frown}{PQ}$ are equal, we have $\angle BMR = \angle RMQ = \angle QMP$, and in the smaller semi-circle, we have $\angle QMP = \angle QBP$. We now consider triangles $\triangle QDB$ and $\triangle QBC$. These triangles have a common side $QB$, and since $MB$ is the diameter of the small semi-circle, we have $\angle DQB = \angle MQB = 90°$, meaning triangles $\triangle QDB$ and $\triangle QBC$ both have right angles in vertex $Q$. Since $\angle QBD = \angle QBP = \angle QMP$, we have $\angle BDQ = 90° - \angle QBD = 90° - \angle QMP$. We know that $\angle BMQ = 2\angle QMP$, and since triangle $MBC$ is isosceles with $MB = MC$, we have $\angle QCB = \angle MCB = 90° - \frac{1}{2} \cdot \angle BMC = 90° - \frac{1}{2} \cdot \angle BMQ = 90° - \angle QMP$. This shows us that $\angle QCB = 90° - \angle QMP = \angle BDQ$. Therefore, we have $\triangle QDB \cong \triangle QBC$ so that $QD = QC$, and $Q$ is then the midpoint of $CD$.

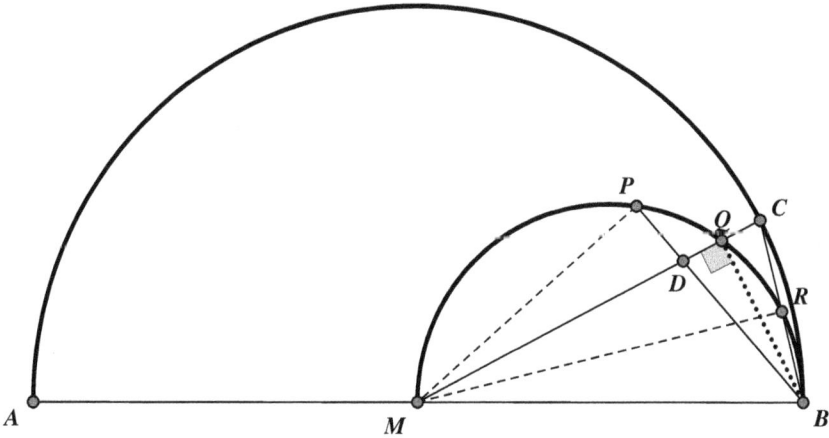

Figure 83-P

## Curiosity 84. A Circle-Related Point with a Double Function

We begin by drawing lines $PR$, $PS$, and $PT$ in Figure 84-P. Since $RB = RC$ are radii of circle $R$, triangle $RCB$ is isosceles with base $BC$. With point $P$ external to circle $R$, the tangents $PB$ and $PC$ are also equal, and triangle $PBC$ is also isosceles with base $BC$. Therefore, line $PR$ is the common altitude of isosceles triangles $RCB$ and $PCB$, and thus the perpendicular bisector of $BC$ and the angle bisector of $\angle CRB = \angle SRT$. Analogously, $PS$ is both the perpendicular bisector of $CA$ and the angle bisector of $\angle ASC = \angle TSR$, and $PT$ is both the perpendicular bisector of $AB$ and the angle bisector of $\angle BTA = \angle RTS$. Point $R$ is both the common point of the perpendicular bisectors of the sides of $\triangle ABC$ and of the angle bisectors of $\triangle RST$. In other words, point $P$ is both the circumcenter of $\triangle ABC$ and the incenter of $\triangle RST$.

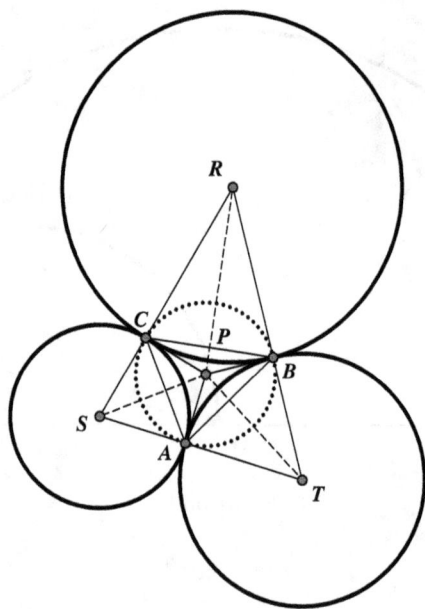

Figure 84-P

## Curiosity 85. Three Circles Through Two Common Points Generate a Surprising Circle

Recall that an angle formed by a tangent and a chord of a circle is equal to one-half the intercepted arc. In Figure 85-P, where $QE$ is tangent to circle $Y$, $\angle AEQ = \frac{1}{2}\stackrel{\frown}{AE} = \angle ABE$. Similarly, since $QF$ is tangent to circle $Z$, $\angle QFA = \frac{1}{2}\stackrel{\frown}{AF} = \angle FBA$. Noting that $\angle ABE + \angle FBA = \angle FBE$, we now consider triangle $QEF$. Here, we note that $\angle EQF = 180° - \angle FEQ - \angle QFE = 180° - \angle AEQ - \angle QFA = 180° - (\angle ABE + \angle FBA) = 180° - \angle FBE$.

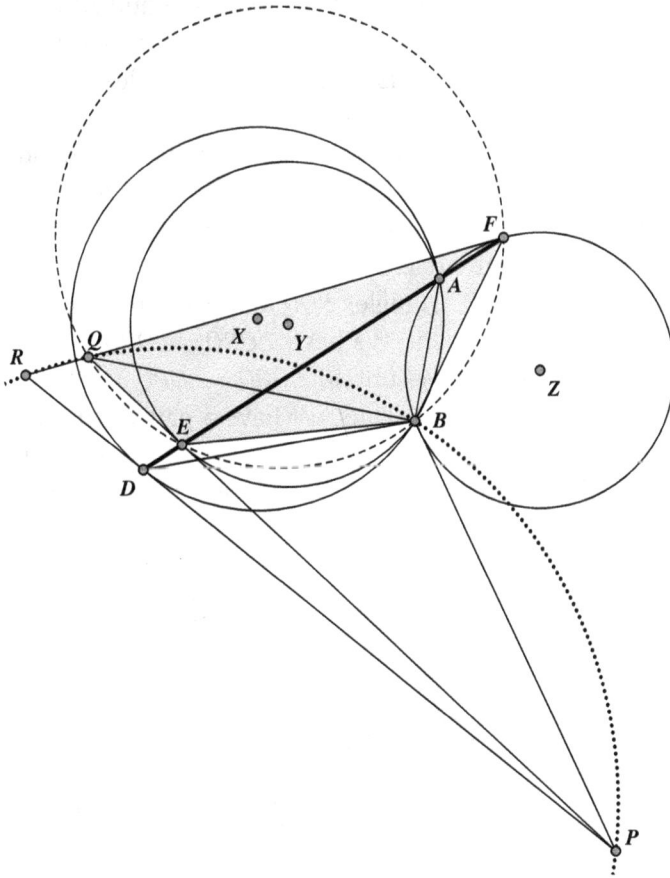

Figure 85-P

Since its opposite angles are supplementary, quadrilateral *FQEB* is cyclic, so that ∠*BFE* = ∠*BQE* because they are both measured by one-half arc *BE*. Analogously, we can show that quadrilateral *FRDB* is also cyclic with ∠*BFD* = ∠*BRD*. Thus, we have ∠*PQB* = ∠*EQB* = ∠*EFB* = ∠*DFB* = ∠*DRB* = ∠*PRB*, which determines quadrilateral *PBQR* to also be cyclic, as these two equal angles are inscribed in the same arc *PB*.

## Curiosity 86. Intersecting Circles Create Five Concyclic Points

In Figure 86-P, we first note that triangles Δ*OPA* and Δ*QBP* are both isosceles, because their sides *OA* = *OP* and *QB* = *QP* are radii of circles *O* and *Q*, respectively. This gives us equal angles ∠*OAP* = ∠*APO* and ∠*PBQ* = ∠*QPB*, and along with the vertical angles ∠*APO* = ∠*QPB*, we get ∠*OAP* = ∠*PBQ*. This allows us to establish the quadrilateral *OQBA* as a cyclic quadrilateral, since both equal angles will be one-half the measure of the same arc *OQ*. We now need to show that point *R* also lies on that circle. Since line *OQ* is a common line of symmetry of circles *O* and *Q*, triangles *POQ* and *RQO* are symmetric with respect to *OQ* and congruent. We have ∠*QRO* = ∠*OPQ*, and considering isosceles triangle *OPA*, we obtain ∠*QRO* = ∠*OPQ* = 180° − ∠*APO* = 180° − ∠*OAP* = 180° − ∠*OAQ*. Thus, we have a pair of supplementary angles ∠*OAQ* and ∠*ORQ* in quadrilateral *AORQ*, which makes it a cyclic quadrilateral. Hence, the points *A*, *O*, *R*, *Q*, and *B* are all on the same circle.

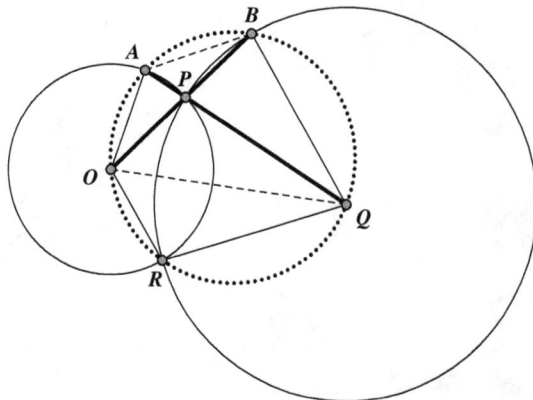

Figure 86-P

## Curiosity 87. Concyclic Centers of Three Related Circumcircles

The goal here is to prove that quadrilateral *PRTS* is cyclic, as shown in Figure 87-P. With point *D* as the intersection of *RT* with *PA* and point *E* as the intersection of *TS* extended with *PC*, we wish to show that $\angle TRP = \angle ESP$. Since *TR* connects the circumcenters of triangles *PAB* and *PBC*, *TR* is the perpendicular bisector of *PA*. We then have *RDT* as the angle bisector in isosceles triangle *AR*, so that $\angle TRP = \frac{1}{2} \cdot \angle ARP$. Similarly, since *TSE* connects the circumcenters of triangles *PAC* and *PCA*, *TSE* is the perpendicular bisector of *PC*, and $\angle ESP = \frac{1}{2} \angle CSP$. In circle *S*, we have the inscribed angle *CBP* and the central angle *CSP* so that $\angle CSP = 2\angle CBP$. In circle *R*, we have $\angle ARP = 2(180° - \angle PBA) = 2\angle CBP$. In summary, this gives us $\angle TRP = \angle DRP = \frac{1}{2}\angle ARP = \angle CBP = \frac{1}{2}\angle CSP = \angle ESP$, or $\angle TRP = \angle ESP$. Quadrilateral *PRTS* is thus cyclic.

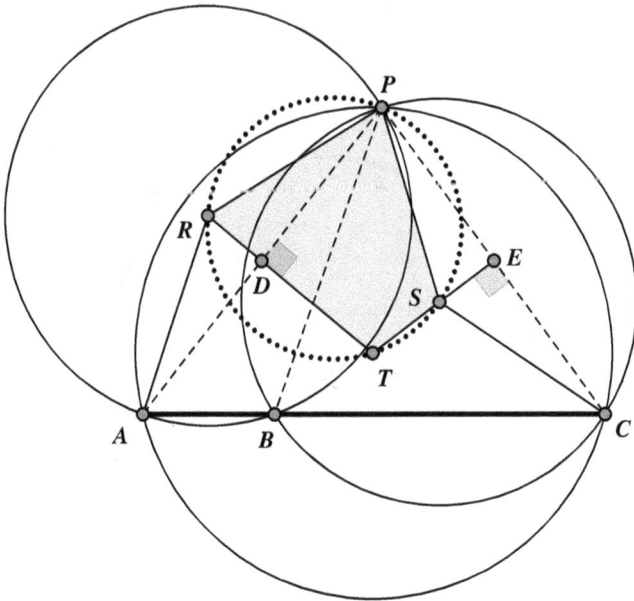

Figure 87-P

## Curiosity 88. Two Unequal Circles Counterintuitively Produce Two Equal Circles

We begin by drawing lines *PA*, *PB*, *PR*, *AR*, and *BR*, as shown in Figure 88-P. In circle *O*, we have $\angle PAB = \angle PRA$, since both angles have one-half the measure of arc *AP*. Similarly, in circle *Q*, we have $\angle ABP = \angle BRP$, as they are both measured by one-half arc *PB*. In triangle *PBA*, we have $\angle PAB + \angle ABP + \angle BAP = 180°$, and we obtain $\angle PRA + \angle BRP + \angle BPA = 180°$, or $\angle BRA + \angle BPA = 180°$. This allows us to get $\sin\angle BRA = \sin\angle BPA$.

We now use a formula for calculating the radius of a circle given the length of a chord and the angle subtended on that chord. If we find the length of chord *AB* of a circle *c*, as shown in Figure 88a-P, as well as the angle $\angle APB$ subtended on arc *AB*, we can calculate the radius *r* of circle *c* in the following way. If point *M* is the midpoint of chord *AB*, $\triangle MBO$ is a right triangle, and since $\angle MOB = \frac{1}{2} \cdot \angle AOB = \frac{1}{2} \cdot 2 \cdot \angle APB = \angle APB$, we obtain $MB = OB \cdot \sin\angle MOB$, or $\frac{1}{2} \cdot AB = r \cdot \sin\angle APB$, which gives us $r = \frac{AB}{2 \cdot \sin\angle APB}$. It is worthwhile to note that $\angle BQA = 180° - \angle APQ$ implies $\sin\angle BQA = \sin\angle APB$, and we can also write $r = \frac{AB}{2 \cdot \sin\angle AQB}$ for a point *Q* on the arc *AB* opposite point *P*.

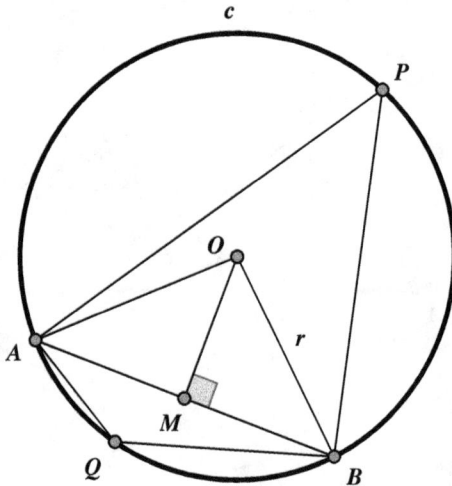

Figure 88a-P

We now return to Figure 88b-P. Noting that the radius of the circumcircle of triangle $APB$ is equal to $\frac{AB}{2\sin \angle BPA}$ and the radius of the circumcircle of triangle $ARB$ is equal to $\frac{AB}{2\sin \angle BRA}$, we see that these radii are equal, and the two circles are of equal size.

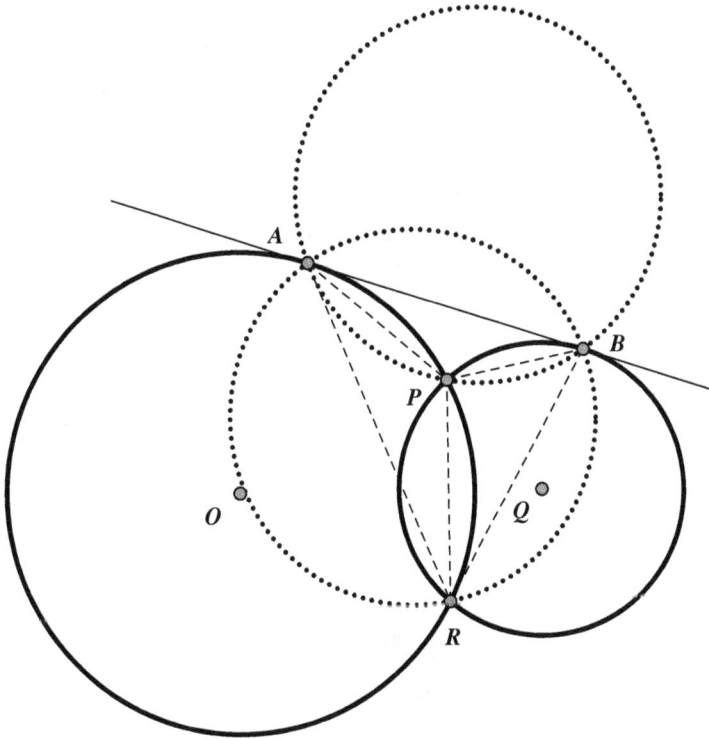

Figure 88b-P

## Curiosity 89. An Angle Property of the Miquel Point of a Triangle

We first consider the problem where point $M$ is in the interior of triangle $ABC$, as shown in Figure 89a-P. Points $D$, $E$, and $F$ are any points on the triangle sides $BC$, $CA$, and $AB$, respectively. Points $P$, $Q$, and $R$ are the circumcenters of triangles $AFE$, $BDF$, and $CED$, respectively. Circles $Q$ and $R$ meet at point $M$, and we draw line segments $MD$, $ME$, and $MF$. Because quadrilateral $BDMF$ is cyclic, and the opposite angles of a cyclic quadrilateral are supplementary, we have $\angle FMC = 180° - \angle DBF = 180° - \angle CBA$. Similarly, quadrilateral $CEMD$ is also cyclic, and we have $\angle DME = 180° - \angle ACB$. When we add these two angles, we get $\angle FMD + \angle DME = 360° - (\angle CBA + \angle ACB)$, or $\angle CBA + \angle ACB = 360° - \angle FMD + \angle DME = \angle EMF$. In triangle $ABC$, we notice that $\angle CBA + \angle ABC = 180° - \angle BAC$, and it follows that $\angle EMF = 180° - \angle BAC$, and we conclude that quadrilateral $AFME$ is cyclic. Therefore, the circle containing points $A$, $E$, and $F$ also contains the common point $M$ of the other two circles.

The angle equality $\angle BFM = \angle AEM$ is because each of these angles is supplementary to $\angle MFA$ in circle $P$. An analogous argument can be made for the other angles, and we obtain $\angle BFM = \angle AEM = \angle CDM$.

Next, we consider the case where point $M$ is outside triangle $ABC$, as shown in Figure 89b-P. Once again, $M$ is the intersection point circles $Q$ and $R$. As quadrilateral $BDMF$ is cyclic, we have $\angle FMD = 180° - \angle DBF = 180° - \angle CBA$, and since quadrilateral $CDME$ is cyclic, we have $\angle EMD = 180° - \angle DCE = \angle ACB$. Subtraction gives us $\angle FME = \angle FMD - \angle EMD = 180° - \angle CBA - \angle ACB = \angle BAC$, and quadrilateral $AEMF$ is therefore also cyclic. Thus, the circle with center $P$ shares the point $M$ with the other two circles in this case as well. Furthermore, the angle equality $\angle MFX = \angle MAE = \angle MDC$ is an immediate consequence of the supplementary angles in circles $P$, $Q$, and $R$, as in the previous case.

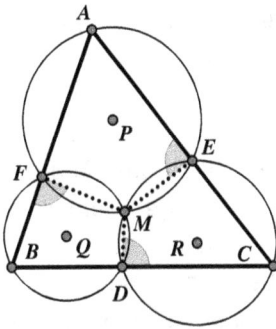

Figure 89a-P

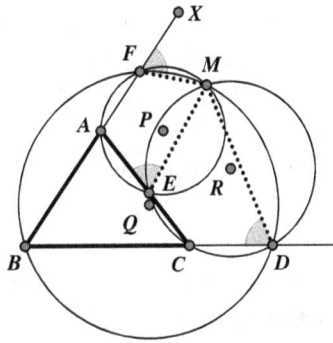

Figure 89b-P

## Curiosity 90. Yet More on the Miquel Relationship

In Figure 90-P, we define point $M$ as the second intersection (besides point $E$) of the circumcircles $R$ and $S$ of triangles $CDE$ and $ACF$, respectively First, we wish to show that this point $M$ is also contained on the circumcircles $P$ and $Q$ of triangles $ADB$ and $BEF$, respectively. In order to do this, we consider the intersection point $X$ of circle $R$ with $FM$ extended. Since $CDXM$ is a cyclic quadrilateral, we have $\angle MCA = \angle MXD$ because both angles are supplementary to $\angle DCM$. Also, since $ACMF$ is a cyclic quadrilateral, we have $\angle MCA + \angle AFM = 180°$. This gives us $\angle MXD + \angle AFM = 180°$ or $\angle FXD + \angle AFX = 180°$, which implies $FA||XD$.

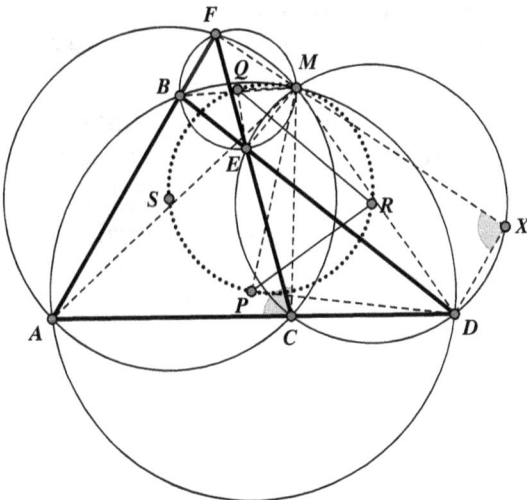

Figure 90-P

As a next step, we apply a similar argument in the opposite direction. Since *EDXM* is a cyclic quadrilateral, we have $\angle FME = \angle XDE$, and since we know that $FA||XD$, we also have $FB||XD$. This gives us $\angle DBF + \angle XDB = 180°$, and thus $\angle DBF + \angle XDE = 180°$, or $\angle EBF + \angle FME = 180°$, which establishes that point *M* lies on the circumcircle *Q* of triangle *BEF*. An analogous argument shows that point *M* also lies on the circumcircle *P* of triangle *ADB*.

Now, we show that the centers of the four circles, *P*, *Q*, *R*, and *S*, all lie on the same circle along with point *M*. We note that in circle *P* we have $\angle DBM = \angle DAM$, and in circle Q we have $\angle EBM = \frac{1}{2}\angle EOM$. Since *ME* is the common chord of circles *Q* and *R*, we then have $\angle RQM = \frac{1}{2}\angle EQM = \angle EBM = \angle DBM$. Similarly, since *MD* is the common chord of circles *P* and *R*, we can show that $\angle RPM = \frac{1}{2}\angle DPM = \angle DAM$. We know that in circle *P* $\angle DBM = \angle DAM$, and this gives us $\angle RPM = DAM = \angle DBM = \angle RQM$; points *M*, *Q*, *P*, and *R* lie on a common circle. In an analogous way, we can also show that points *M*, *S*, *P*, and *R* lie on a common circle, and we see that all five points *M*, *P*, *Q*, *R*, and *S* lie on a common circle.

## Curiosity 91. Three Intersecting Circles Generate Similar Triangles and More

We begin by drawing the common chord *OR* of circles *A* and *C* in Figure 91-P. Line *AC* is the perpendicular bisector of *OR*. If point *M* is the intersection of *AC* with circle *A*, then $\angle ODR = \frac{1}{2}\cdot\angle OAR = \angle OAM = \angle OAC$; therefore, $DR||AC$. Analogously, in circle *C* we have $FR||AC$, and *FRD* is thus a straight line parallel to *AC*. The same can be done to show that *DPE* and *EQF* are also straight lines and parallel to *AB* and *BC*, respectively. Therefore, we have similar triangles *ABC* and *DEF*, with the sides of triangle *DEF* containing the intersection points *P*, *Q*, and *R*.

This Curiosity follows immediately from an elementary application of homothety (see Toolbox). Because point $A$ is the midpoint of $OD$, point $B$ is the midpoint of $OE$, and point $C$ is the midpoint of $OF$, triangle $DEF$ results from applying a homothety with center $O$ and factor 2 to triangle $ABC$. The two triangles are similar. Furthermore, since $CA$ is the bisector of $OR$, as shown above, and $AB$ and $BC$ are the bisectors of $OP$ and $OQ$, respectively, points $P$, $Q$, and $R$ must lie on the sides of triangle $DEF$.

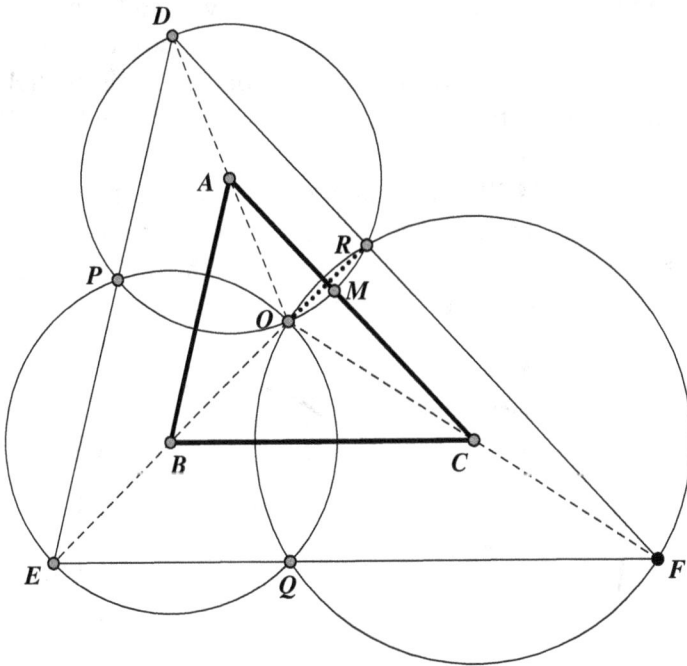

Figure 91-P

## Curiosity 92. Equal Circles Generated by Three Equal Circles with a Common Point

In Figure 92-P, we first note that we have $PQ = PR = PS$, since point $P$ is the point of intersection of circles $Q$, $R$ and $S$, which have equal radii. This implies that points $Q$, $R$ and $S$ all lie on a common circle with center $P$, and the same radius as the three original circles. We see that $AS = SB = BQ = QC = CR = RA = PQ = PR = PS$, which results in the hexagon $ASBQCR$, which has six sides of equal length. Since quadrilaterals $ASPR$ and $BQPS$ are both rhombi, their opposite sides are parallel, and we have $AR||SP||BQ$. Similarly, we also have $BS||QP||CR$ and $CQ||RP||AS$. All pairs of opposite sides in $ASBQCR$ are, therefore, parallel. We let point $D$ be the intersection of the line through $A$ parallel to $BS$ and the line through $B$ parallel to $AS$. Since we have $AS = SB$, quadrilateral $ASBD$ is also a rhombus. If we now consider quadrilateral $ADCR$, we have $AD||SB||RC$ and $CR = RA = AD$. It follows that $ADCR$ is also a rhombus, and we have $DC = DA = DB$. Point $D$ is then the circumcenter of triangle $ABC$, with $DA = SA$, and the radius of circle $D$ is equal to the radius of circle $S$, and equal to all the other radii.

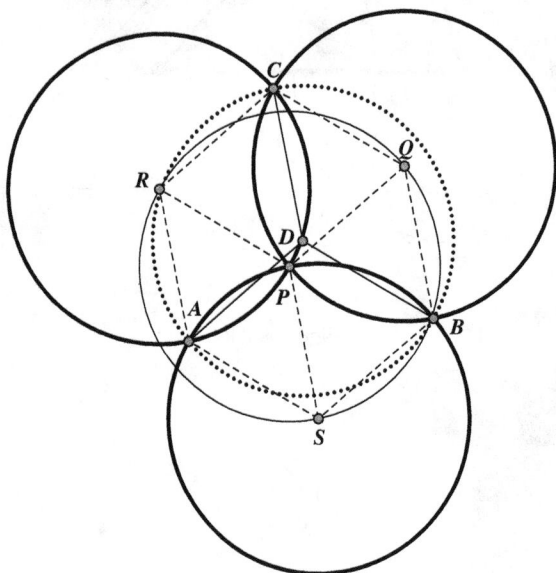

Figure 92-P

## Curiosity 93. Congruent Figures Generated by Three Equal Circles with a Common Point

In Figure 93-P, we first consider quadrilateral *ASQC*. From the proof of Curiosity 92, we know that *AS*||*CQ* and *AS* = *QC*. This establishes *ASQC* as a parallelogram so that *AC* = *SQ* and *AC*||*SQ*. Similarly, in parallelogram *APQD*, we obtain *PA* = *QD* and *PA*||*QD*; in parallelogram *BDRP*, we obtain *BP* = *DR* and *BP*||*DR*; and in parallelogram *CRSB*, we obtain *CB* = *RS* and *CB*||*RS*. We see that corresponding sides of the two quadrilaterals *APBC* and *QDRS* are equal and parallel, which implies that the quadrilaterals are congruent.

We need to prove that the line joining any two points in one of these quadrilaterals is perpendicular to the line joining its other two points. Let *N* denote the intersection of *BC* with *AP* extended. We have already established *PA*||*QD*, and since *BC* is the common chord of circles *D* and *Q*, we have *DQ* ⊥ *BC*, so that *APN* ⊥ *BC* as well. In a similar way, we can show that *BP* ⊥ *AC*, and since quadrilaterals *APBC* and *QDRS* are congruent, we also have *DR* ⊥ *QS* and *DQ* ⊥ *RS*.

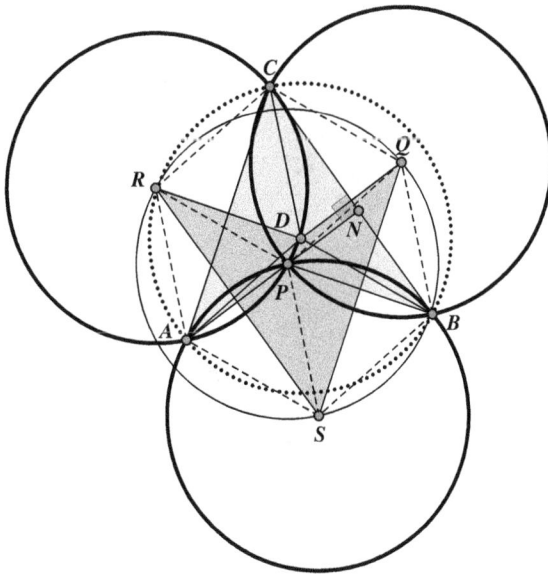

Figure 93-P

## Curiosity 94. Fun with Intersecting Circles

In Figure 94-P, quadrilaterals $ABKN$ and $BCLK$ are both cyclic, so that we obtain $180° - \angle NKB = \angle BAN$ and $180° - \angle BKL = \angle LCB$, and thus $\angle LKN = 360° - \angle NKB - \angle BKL = (180° - \angle NKB) + (180° - \angle BKL) = \angle BAN + \angle LCB$. Similarly, since quadrilaterals $CDML$ and $DANM$ are both cyclic, we obtain $180° - \angle LMD = \angle DCL$ and $180° - \angle DMN = \angle NAD$, and thus $\angle NML = 360° - \angle LMD - \angle DMN = (180° - \angle LMD) + (180° - \angle DMN) = \angle DCL + \angle NAD$. Adding these equations gives us

$$\begin{aligned}
\angle LKN &+ \angle NML \\
&= (\angle BAN + \angle LCB) + (\angle DCL + \angle NAD) \\
&= (\angle BAN + \angle NAD) + (\angle DCL + \angle LCB) \\
&= \angle BAD + \angle DCB.
\end{aligned}$$

Since $ABCD$ is a cyclic quadrilateral, we have $\angle BAD + \angle DCB = 180°$, which gives us $\angle LKN + \angle NML = 180°$, and quadrilateral $KLMN$ is therefore also cyclic.

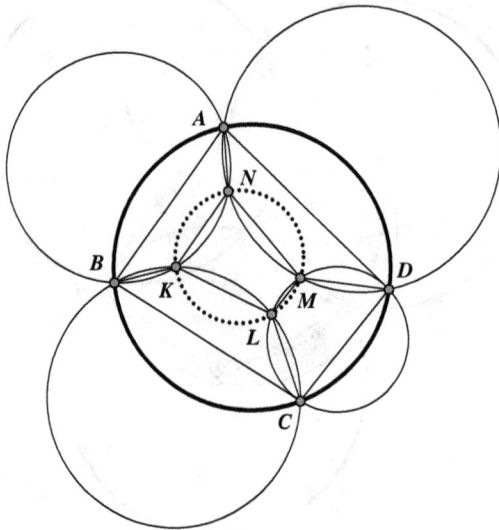

Figure 94-P

## Curiosity 95. Four Equal Circles Generate a Circumscribable Quadrilateral

To Figure 95-P, we have added and named the points of tangency of the four circles with their outside tangents. Furthermore, we have added the radii to these points of tangency and the sides of quadrilateral *ABCD*.

We first consider quadrilateral $AA''EA'$. Since the radii $AA'$ and $AA''$ are perpendicular to the tangents $EF$ and $HE$, respectively, we have $\angle EA''A = \angle AA'E = 90$. Thus, $\angle A''AA' = 180° - \angle A'EA'' = 180° - \angle FEH$. Noting that the circles all have equal radii, and the radii are all perpendicular to the respective tangents, we see that quadrilaterals $AA'B''B$ and $ADD'A''$ are rectangles, and we have $\angle A'AB = \angle DAA'' = 90°$. From this, we obtain $\angle BAD = 360° - \angle A'AB - \angle A''AA' - \angle DAA'' = 360° - 90° - (180° - \angle FEH) - 90° = \angle FEH$. Analogously, we can also consider quadrilateral $CC''GC'$ to obtain $\angle DCB = \angle HGF$. All that remains is to notice that quadrilateral *ABCD* is cyclic, as all points *A*, *B*, *C*, and *D* are the same distance from the intersection point *P* of the four circles. This gives us $\angle FEH + \angle HGF = \angle BAD + \angle DCB = 180°$, and since $\angle FEH + \angle HGF = 180°$, the four points *E*, *F*, *G*, and *H* are concyclic.

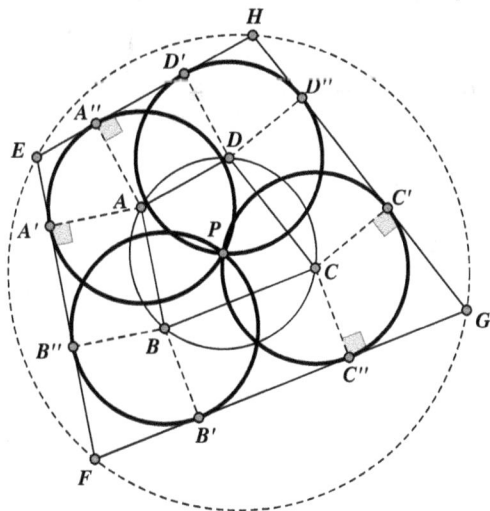

Figure 95-P

## Curiosity 96. The Reuleaux Triangle

As we can see in Figure 96-P, drawing a tangent at point $P$ to the arc of the circle with center $A$, the distance from point $P$ to the circle's center $A$ is equal to the radius of the circle. This radius is equal to the lengths $b = AB = CA$ of two sides of equilateral triangle $ABC$, as these are also radii of the circle. The perpendicular to $AP$ at point $A$ is such that $AP = b$. This is also true for the tangent to the arc of the circle with center $C$ at point $Q$, as all sides of the equilateral triangle have the length $b$. Since an analogous argument holds for tangents of the third arc, we see that pairs of parallel tangent lines of the figure always have a distance equal to the length $b$ of the sides of triangle $ABC$.

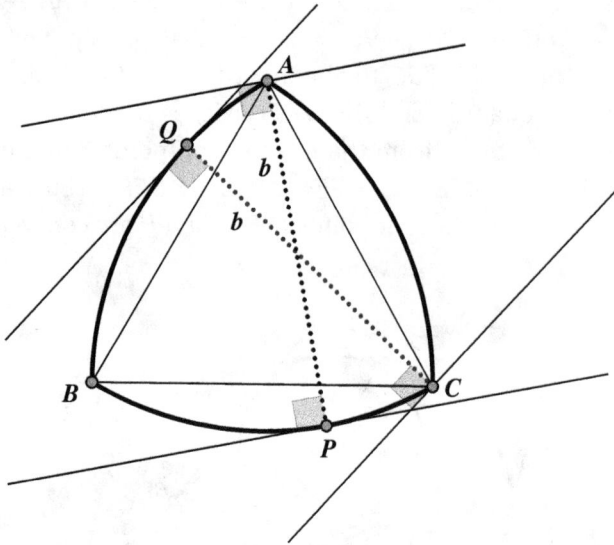

Figure 96-P

## Curiosity 97. Some Other Shapes with Constant Breadth

The same argument we used in explaining Curiosity 96 will also work here. We first consider the left-hand diagram in Figure 97-P, which is based on a regular pentagon. The radius of the circle with center $A$ and a tangent at point $P$ is equal to the lengths $d = AC = AD$ of two sides isosceles triangle $ACD$, which are also radii of the circle. This length $d$ is equal to the length of the diagonals in pentagon $ABCDE$, and equal to the distance between the tangent line of the figure through $A$ and the tangent at $P$. This is true if point $P$ is chosen on any of the other arcs of the circle comprising the figure. Similarly, considering the right-hand figure, which is based on the regular heptagon, the argument can be repeated in exactly the same way for the tangent at point $Q$ on the arc of the circle with center $F$ and the tangent at point $F$, with the length $e$ of the long diagonal of the heptagon yielding the common distance of all such pairs.

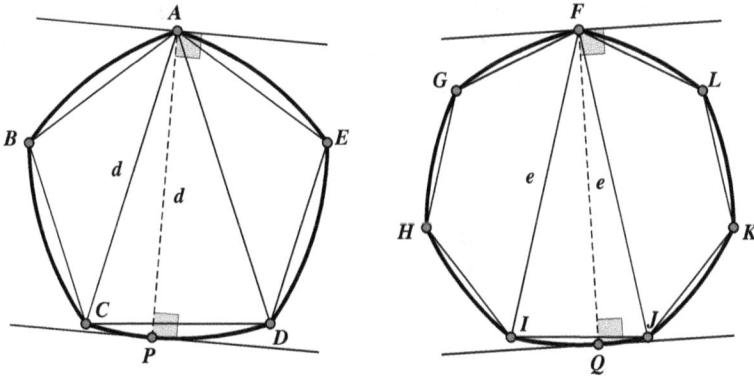

Figure 97-P

## Curiosity 98. Introducing the Radical Axis of Intersecting Circles

Recall that when a secant and a tangent segment are drawn to the same circle from the same external point, the tangent segment is the mean proportional between the secant and its external segment. We see this illustrated in Figure 98-P. With respect to circle $O$, we have $PR$ as the mean proportional between $PA$ and $PB$, and we have $PR^2 = PA \cdot PB$. Similarly, with respect to circle $Q$, we have $PS$ as the mean proportional between $PA$ and $PB$, so that $PS^2 = PA \cdot PB$. For any external point on the line $AB$, we have $PR = PS$.

Now, suppose that $P$ is any point in the plane for which for the tangent segments $PR$ and $PS$ to circles $O$ and $Q$, respectively, yield $PR = PS$. Let $PA$ intersect circle $O$ at point $B$ and circle $Q$ at point $B'$. As before, we then have $PR^2 = PA \cdot PB$ and $PQ^2 = PA \cdot PB'$. Since $PR = PS$, we then have $PB = PB'$, and thus, $B = B'$. Point $P$, therefore, lies on the common secant $AB$ of the two circles.

Finally, choosing point $T$ on secant $AB$, as shown in Figure 98-P, we find that $TU \cdot TV = TW \cdot TX$ holds for any selected lines $TUV$ and $TWX$, as the tangents from $T$ to the two circles are of equal length.

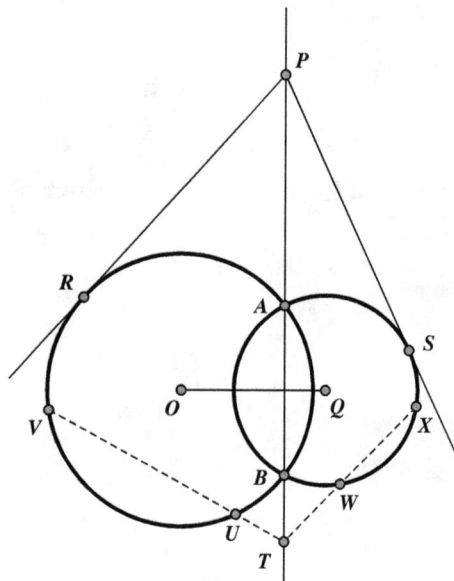

Figure 98-P

## Curiosity 99. The Radical Axis of Tangent Circles

The situations illustrated in Figure 99a-P and 99b-P are a bit simpler than that described in Curiosity 98. Here, it is clear that *PZ* has the same length with respect to both circles for any point chosen on their common tangent in *Z*.

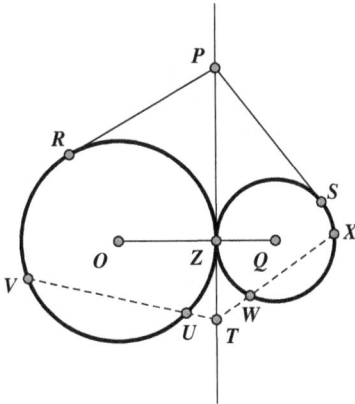 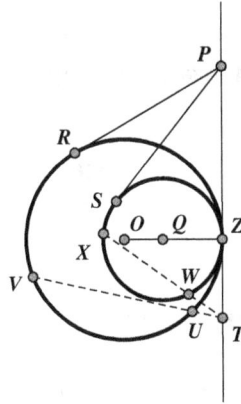

Figure 99a-P                          Figure 99b-P

If *P* is any point in the plane for which *PR* = *PS* holds true for the tangent segments *PR* and *PS* to circles *O* and *Q*, respectively, we let *PZ* intersect circle *O* at point *B* and circle *Q* at point *B'*, as illustrated in Figure 99c-P. (Note that this only shows the case of externally tangent circles. The argument for internally tangent circles is identical.) Again, we then have $PR^2 = PZ \cdot PB$ and $PQ^2 = PZ \cdot PB'$, and thus, *B* = *B'*. This is

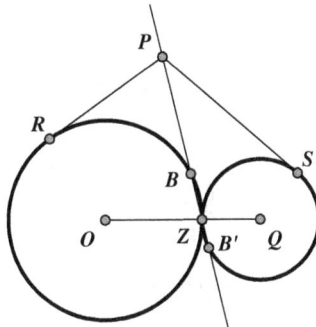

Figure 99c-P

only possible for $B = B' = Z$, and point $P$ therefore lies on the common tangent of the two circles at point $Z$.

Finally, as in the proof of Curiosity 98, for point $T$ on the common tangent in $Z$, as shown in the figures, $TU \cdot TV = TW \cdot TX$ holds true for any choice of lines $TUV$ and $TWX$, as the tangents from $T$ to the two circles are of equal length.

## Curiosity 100. The Radical Axis of External Non-Intersecting Circles

Although this Curiosity is closely related to Curiosities 98 and 99, we will find it much easier to prove by applying analytic geometry methods. Perhaps this can serve as an instructive example of how analytic geometry can sometimes be more efficient than synthetic geometry.

As we see in Figure 100-P, we choose a system of coordinates in such a way that point $O$ is in the origin, with coordinates $(0,0)$. Point $Q$ lies on the positive side of the $x$-axis, and the coordinates of $Q$ are $(m,0)$ for some positive number $m$. The point $P$, with coordinates $(x,y)$, has the property that the tangents $PR$ and $PS$ from $P$ to circles $O$ and $Q$, respectively, are congruent. Finally, we have $r$ as the radius of circle $O$, and $s$ as the radius of circle $Q$.

We are given $PR = PS$, which is equivalent to $PR^2 = PS^2$. Since the radius of circle $O$ through point $R$ is perpendicular to the tangent $PR$, we have $\triangle PRO$ as a right triangle, with $PR^2 = OP^2 - OR^2 = (x^2 + y^2) - r^2$. Similarly, $\triangle PQS$ is also a right triangle, with $PS^2 = QP^2 - QR^2 = ((x - m)^2 + y^2) - s^2$. Substituting these expressions gives us $x^2 + y^2 - r^2 = x^2 - 2mx + m^2 + y^2 - s^2$, which simplifies to $2mx = m^2 + r^2 - s^2$ or $x = \frac{m^2 + r^2 - s^2}{2m}$. This means that any point $P$ with the given property has this specific $x$-coordinate, and thus lies on a line perpendicular to the $x$-axis, which is the line $OQ$. We can also note that any choice of $y$-coordinate will give us a point $P$ with $PR = PS$, as we can simply do the calculation in the opposite direction. The radical axis of the two circles is a line perpendicular to $OQ$. Finally, we note that the equality $TU \cdot TV = TW \cdot TX$ holds true for all points $T$ on this line, and all secants $TUV$ of circle $O$ and all secants $TWX$ of circle $Q$ follow in the same way as in the proofs of Curiosities 98 and 99.

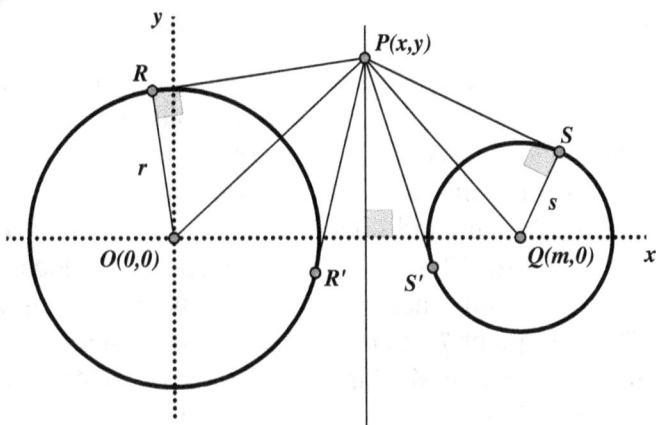

Figure 100-P

## Curiosity 101. The Radical Axis of Internal Non-Intersecting Circles

The proof of this Curiosity is identical to that of Curiosity 100, and uses the system of coordinates shown in Figure 101-P.

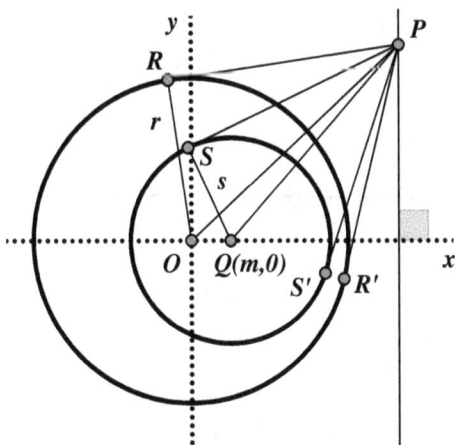

Figure 101-P

## Curiosity 102. A Coincidence of Intersecting Circles

As we know from Curiosity 98, a point $P$ lies on the radical axis $KL$ of circles $O$ and $Q$ if and only if $PU \cdot PV = PW \cdot PX$ holds true for any secant $PUV$ of circle $O$ and any secant $PWX$ of circle $Q$. In order to show that the orthocenter $H$ has this property, we consider the altitudes $AS$ and $BT$ of triangle $ABC$, with $S$ and $T$ denoting the feet of the altitudes from $A$ and $B$, respectively. This situation is illustrated in Figure 102-P. Since $\angle FSA = 90°$, point $S$ lies on the circle $O$ with diameter $AF$, and since $\angle BTG = 90°$, point $T$ lies on the circle $Q$ with diameter $BG$. The orthocenter $H$ of triangle $ABC$ is the intersection of $AS$ and $BT$. As $\angle ATB = \angle ASB = 90°$, points $S$ and $T$ both lie on the circle with diameter $AB$. With respect to this circle, we obtain $AH \cdot SH = BH \cdot TH$. Since $A$ and $S$ both lie on circle $O$, we can interpret the product $AH \cdot SH$ with respect to circle $O$, and since $B$ and $T$ both lie on circle $Q$, we can interpret the product $BH \cdot TH$ with respect to circle $Q$. Having established these two values to be equal, we see that point $H$ lies on the radical axis $KL$ of circles $O$ and $Q$.

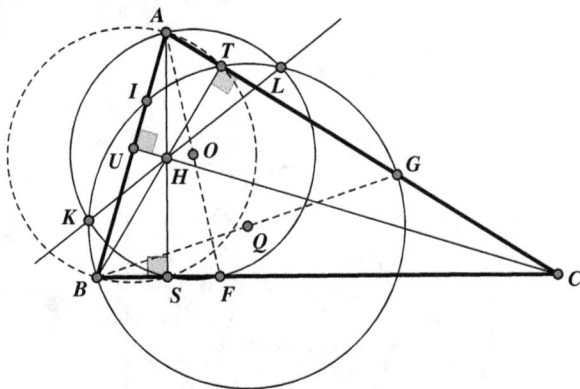

Figure 102-P

## Curiosity 103. The Radical Center of Three Circles

Consider circles $O$, $Q$, and $R$, whose radical axes are shown in Figure 103-P. We have chosen points in the figure such that $AB$ is the radical axis of circles $O$ and $Q$ and $CD$ is the radical axis of circles $O$ and $R$. Let $P$ be the intersection of $AB$ and $CD$. As $P$ lies on the radical axis $AB$ of circles $O$ and $Q$, the equality $PU \cdot PV = PW \cdot PX$ holds true for all secants *PUV* of circle *O* and all secants *PWX* of circle *Q*. As *P* is external to the circles, the tangents from *P* to the circles are of equal length, which we see illustrated in Figure 103-P as *PE* = *PH*. Similarly, as point *P* is on the radical axis *CD* of circles *O* and *R*, the equality $PU \cdot PV = PY \cdot PZ$ holds true for all secants *PUV* of circle *O* and all secants *PYZ* of circle *R*. With *P* external to these circles, the tangents from *P* to the circles are again of equal length, which we see as *PE* = *PJ*. This means that *PE* = *PH* = *PJ*, and that the equality $PU \cdot PV = PW \cdot PX = PY \cdot PZ$ holds true for all secants *PUV* of circle *O*, all secants *PWX* of circle *Q*, and all secants *PYZ* of circle *R*. Point *P* is therefore also on the radical axis of circles *Q* and *R*.

With *P* external to all three circles, we have a total of six tangents from *P* of equal length: *PE*, *PF*, *PG*, *PH*, *PJ*, and *PK*. Considering point *E*, we note that the radius *OE* of circle *O* is perpendicular to the tangent *PE*. Therefore, *EO* is a tangent of the circle *c* with center *P* containing point *E*, and circles *O* and *P* intersect at right angles in *E*. This can be argued in an analogous way for all six tangents, and we see that *c* intersects all three given circles at right angles, thus solving *Monge's problem*. Such a circle only exists if the radical axes of the given circles intersect externally.

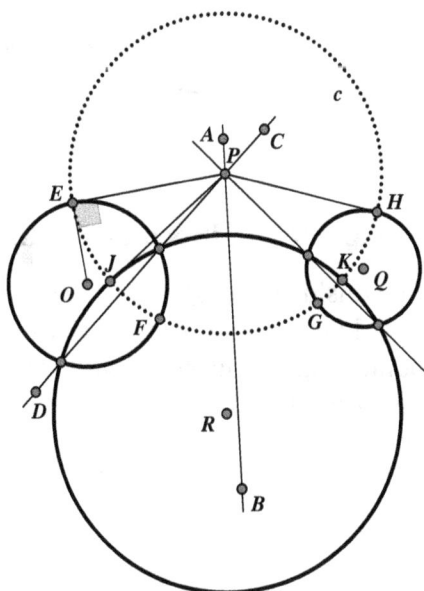

Figure 103-P

## Curiosity 104. A Surprising Result from a Randomly Selected Point Outside a Circle

Since we are given $OP \cdot OP' = r^2$ in Figure 104-P, we can express this as $\frac{OP}{r} = \frac{r}{OP'}$, or $\frac{OP}{OA} = \frac{OA}{OP'}$. This establishes $\triangle OAP \sim \triangle OAP'$ because they also have a common angle at vertex $O$. Mapping points $P$ onto points $P'$ in this way is commonly referred to as *inversion* on the circle $c$.

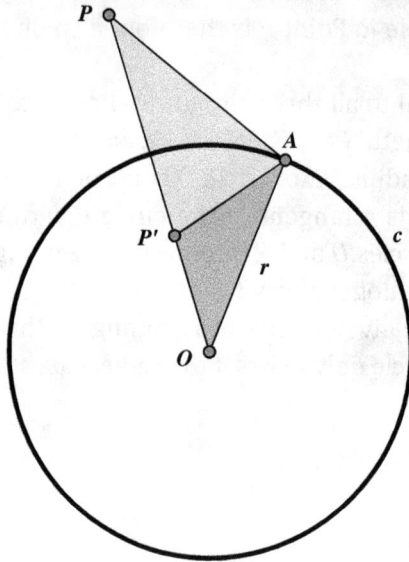

Figure 104-P

## Curiosity 105. Unexpected Points on a Circle's Tangent

In Figure 105-P, we consider triangles $OAP$ and $OAP'$. Since we are given $OP \cdot OP' = r^2$, we know from Curiosity 104 that these triangles are similar. Since point $P$ lies on the circle with diameter $OA$, we have $\angle OPA = 90°$, and thus, $\angle P'AO = 90°$. Therefore, point $P'$ lies on the tangent $t$ of circle $c$ at point $A$.

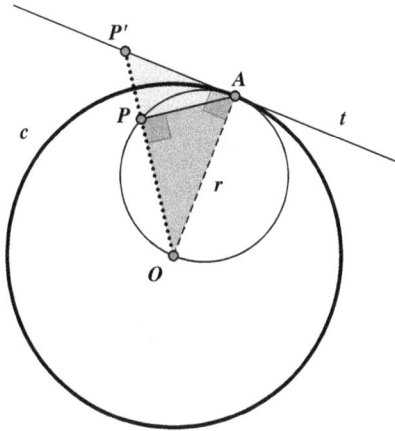

Figure 105-P

## Curiosity 106. Points on a Secant Generate Points on a Special Circumcircle

In Figure 106-P, we consider triangles $OBP$ and $OBP'$. Since we are given $OP \cdot OP' = r^2$, we know from Curiosity 104 that these triangles are similar, and we therefore have $\angle OP'B = \angle PBO = \angle ABO$. As this angle is independent of the choice of point $P$ on $AB$, all resulting points $P'$ lie on the circumcircle of triangle $OBA$. Note that this proof can be extended in the same way to points $P$ on line $s$ outside circle $c$.

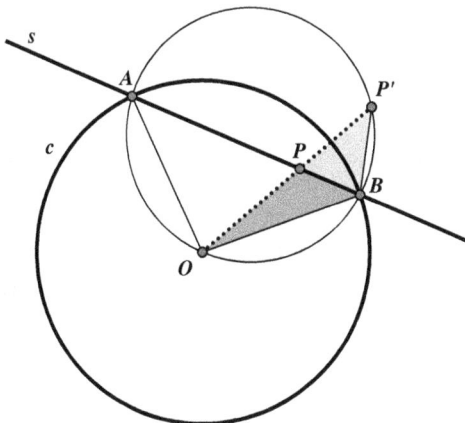

Figure 106-P

## Curiosity 107. An External Circle Generates an Internal Circle

In Figure 107-P, we first draw diameter $AB$ of circle $d$ and extend it to center $O$. We determine points $A'$ and $B'$ such that $OA \cdot OA' = r^2$ and $OB \cdot OB' = r^2$, and draw lines $PA$, $PB$, $P'A'$, and $P'B'$. Since point $P$ lies on a circle with diameter $AB$, we note that $\angle BPA = 90°$. We now consider triangles $\triangle OBP$ and $\triangle OB'P'$. These have a common angle at point $O$, and since $OB \cdot OB' = r^2 = OP \cdot OP'$, we have $\frac{OB}{OP} = \frac{OP'}{OB'}$, and thus, $\triangle OBP \sim \triangle OB'P'$. This gives us $\angle ABP = \angle B'P'P$. Similarly, since $OA \cdot OA' = r^2 = OP \cdot OP'$, we have $\frac{OA}{OP} = \frac{OP'}{OA'}$, and thus, $\triangle OAP \sim \triangle OA'P'$, which gives us $\angle PAO = \angle OP'A'$. We are now ready to consider angle $\angle A'P'B$. Summing up, we have $\angle A'P'B = 180° - \angle OP'A' - \angle B'P'P = 180° - \angle PAO - \angle ABP = \angle BPA$. Because $\angle BPA$ is a right angle, as $AB$ is the diameter of circle $d$, we also have $\angle A'P'B' = 90°$, and we see that all points $P'$ lie on the circle with diameter $A'B'$.

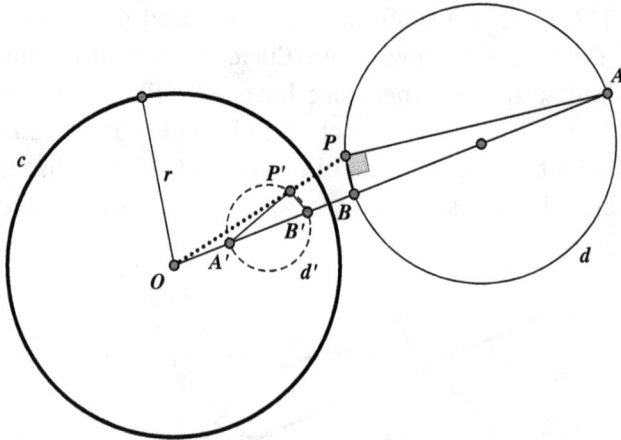

Figure 107-P

## Curiosity 108. Two Circles Generating Concyclic Points

The proof for this Curiosity is essentially the same as that for Curiosity 107. As we did in Figure 107-P, we first draw the diameter $AB$ of circle $d$ that contains center $O$, but in Figure 108-P, we extend this diameter beyond circle $c$. We again determine points $A'$ and $B'$ such that $OA \cdot OA' = r^2$ and $OB \cdot OB' = r^2$, and then draw lines $PA$, $PB$, $P'A'$, and $P'B'$. Since point $P$ lies on a circle with diameter $AB$, it follows that $\angle BPA = 90°$.

We now consider triangles $\triangle OBP$ and $\triangle OB'P'$. These have a common angle at point $O$, and since $OB \cdot OB' = r^2 = OP \cdot OP'$, we can write this as $\frac{OB}{OP} = \frac{OP'}{OB'}$, and thus, $\triangle OBP \sim \triangle OB'P'$, whereupon we have $\angle PBA = \angle OP'B'$. Similarly, since $OA \cdot OA' = r^2 = OP \cdot OP'$, we can write this as $\frac{OA}{OP} = \frac{OP'}{OA'}$, and thus, $\triangle OPA \sim OP'A'$, and it follows that $\angle OAP = \angle A'P'O$. We are now ready to consider angle $\angle A'P'B'$. Summing up, we have $\angle A'P'B' = \angle A'P'O + \angle OP'B' = \angle OAP + \angle PBA = 180° - \angle APB$. Because $\angle APB$ is a right angle, as $AB$ is the diameter of circle $d$, we once again have $\angle A'P'B' = 90°$, and we see that all points $P'$ lie on the circle with diameter $A'B'$. Since the points $P$ were all in the interior of circle $c$, points $P'$ are all external to circle $c$, and circle $c$ is therefore in the interior of the circle with diameter $A'B'$.

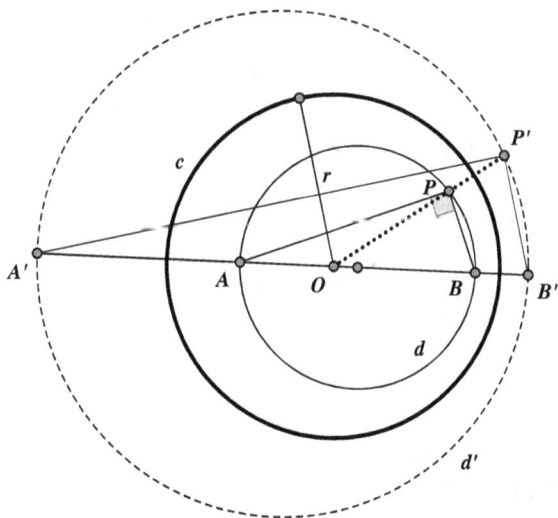

Figure 108-P

## Curiosity 109. Generating Semicircles on the Sides of a Square

In the proof of Curiosity 106, we showed that the points $P$ on chord $AB$ of circle $O$ determine points $P'$ on the circumcircle of triangle $OAB$ external to circle $O$. As we see in Figure 109-P, triangle $OAB$ is a right isosceles triangle because the diagonals $AC$ and $BD$ of the square $ABCD$ are perpendicular and bisect each other. This means that $AB$ is a diameter of this circle, and points $P'$ resulting from points $P$ on $AB$ all lie on a semicircle with diameter $AB$, external to circle $O$. We can argue in the same way for chords $BC$, $CD$, and $DA$, and we will see that the points on the square determine the points of the four semi-circles with diameters $AB$, $BC$, $CD$, and $DA$ external to circle $O$.

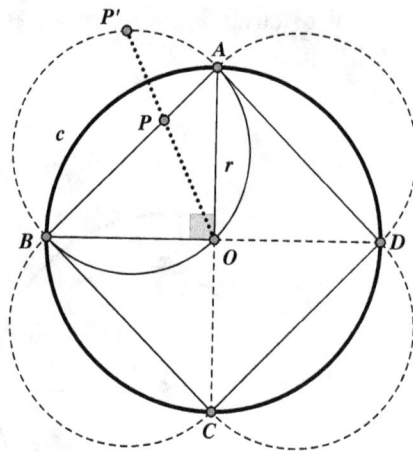

Figure 109-P

## Curiosity 110. Two Circles Generating an Unexpected Equilateral Triangle

In Figure 110-P, we first draw the radii $OD$, $OQ$, $OA$, $QA$, and $QD$. These radii generate two equilateral triangles, $\triangle DOQ$ and $\triangle AQO$, and therefore, $\angle AOD = \angle DQA = 120°$. Since quadrilateral $QDCA$ is cyclic and inscribed in circle $O$, we have $\angle ACB = \angle ACD = 180° - \angle DOA = 180° - 120° = 60°$. Furthermore, since quadrilateral $DOBA$ is cyclic inscribed and in circle $Q$, we have $\angle ABD = \angle AOD = 120°$, which gives us

$\angle CBA = 180° - \angle ABD = 180° - 120° = 60°$. We see that two angles in triangle *ABC* are equal to 60°, and triangle *ABC* is therefore equilateral.

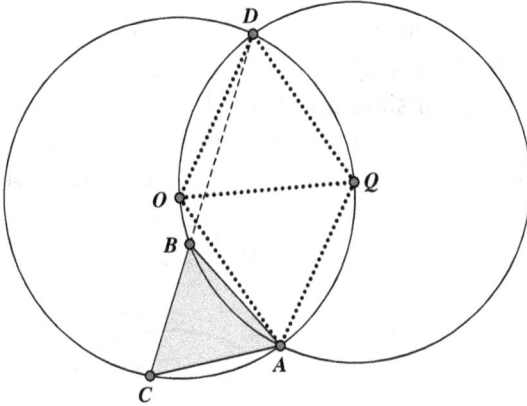

Figure 110-P

## Curiosity 111. Two Equal Circles Produce an Isosceles Triangle

In Figure 111-P, we draw the common chord *AB* of the two circles as well as the radii *OA*, *OB*, *QA*, and *QB*. Since triangles $\triangle OBA$ and $\triangle QAB$ share a side *AB*, and their other sides are of equal length, we have $\triangle OBA \cong \triangle QAB$, and thus $\angle BOA = \angle AQB$. In circle *O*, we have $\angle BXA = \frac{1}{2}\angle BOA$, and in circle *Q*, we have $\angle AYB = \frac{1}{2}\angle AQB$. Therefore, $\angle BXY = \angle BXA = \frac{1}{2}\angle BOA = \frac{1}{2}\angle AQB = \angle AYB = \angle XYB$, and with $\angle BXY = \angle XYB$, we see that $\triangle BYX$ is isosceles.

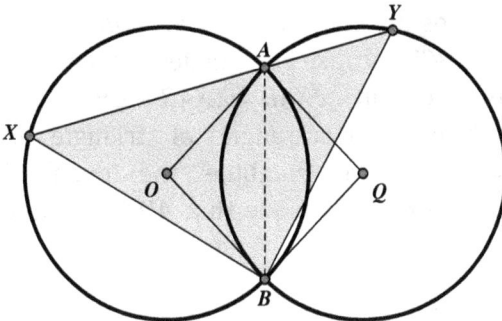

Figure 111-P

## Curiosity 112. Two Equal Circles Generate Another Isosceles Triangle

The proof in this case is almost identical to the proof of Curiosity 111. In Figure 112-P, we have drawn the common chord $AB$ of the two circles and radii $OA$, $OB$, $QA$, and $QB$. Since triangles $\triangle OBA$ and $\triangle QAB$ share a side $AB$, and their other sides are of equal length, we have $\triangle OBA \cong \triangle QAB$ and $\angle BOA = \angle AQB$. In circle $O$, we have $\angle AXB = 180° - \frac{1}{2}\angle BOA$, and $\angle BXY = 180° - \angle AXB = \frac{1}{2}\angle BOA$. In circle $Q$, we have $\angle AYB = \frac{1}{2}\angle AQB$, and this gives us $\angle BXY = \frac{1}{2}\cdot\angle BOA = \frac{1}{2}\cdot\angle AQB = \angle AYB = \angle XYB$. With $\angle BXY = \angle XYB$, we see that $\triangle BYX$ is isosceles.

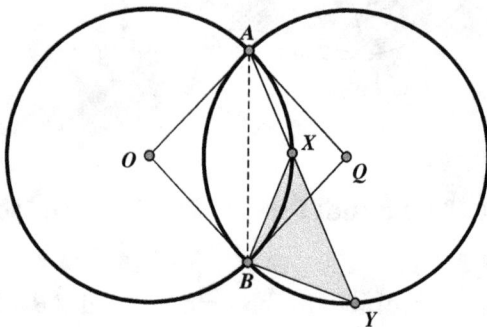

Figure 112-P

## Curiosity 113. An Unexpected Angle Bisector

First, we draw lines $AB$ and $BY$, as shown in Figure 113-P. We first notice that $\angle DAY = \angle DBY$ in circle $Q$ because both angles are measured by arc $DY$. Also in circle $Q$, we have $\angle CYB = \angle XAB = \angle XYB$ because both angles are measured by arc $XB$. In circle $O$, we have $\angle BAE = \angle BCE = \angle BCY$ because both angles are measured by arc $BE$.

We therefore obtain $\angle XAD = \angle XAB + \angle BAE = \angle XYB + \angle BCE = \angle CYB + \angle BCY$. The exterior angle of triangle $BYC$ generates $\angle CYB + \angle BCY = \angle DBY$. We combine the equalities to obtain $\angle XAD = \angle DBY = \angle DAY$, and we see that $AD$ is the angle bisector of $\angle XAY$.

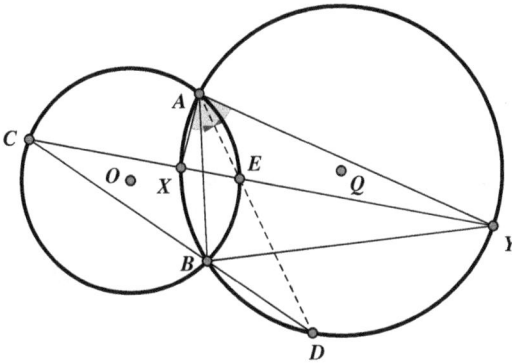

Figure 113-P

## Curiosity 114. An Amazing Angle Equality in Intersecting Circles

In Figure 114-P, when we draw line $XY$, we have $\angle AXC = \angle AXY - \angle CXY$. In circle $O$, we have $\angle AXY = \angle ABY$, since both angles are one-half the measure of arc $AY$. Analogously, in circle $Q$, we have $\angle CXY = \angle CDY$, which gives us $\angle AXC = \angle AXY - \angle CXY = \angle ABY - \angle CDY = \angle ABY - \angle BDY$. In triangle $BYD$, considering exterior angle $ABY$, we have $\angle ABY - \angle BDY = \angle DYB$, and we thus obtain $\angle AXC = \angle DYB$.

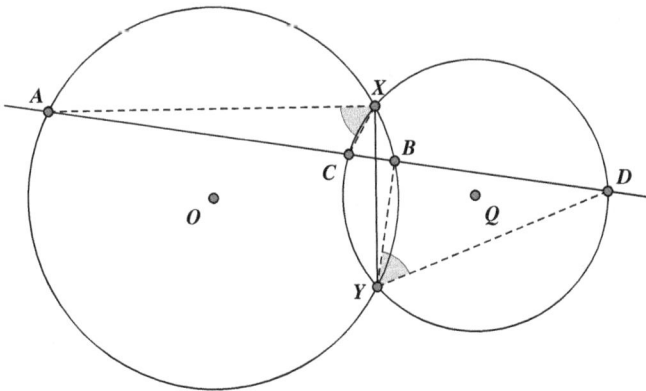

Figure 114-P

## Curiosity 115. A Cyclic Quadrilateral Generates Two Circles with Perpendicular Radii

In Figure 115-P, we begin by drawing the circumcircle of triangle *AEB*, which intersects line *EF* at point *G*, as well as drawing lines *OD*, *OE*, *OF*, and *OM*. Both *ABCD* and *AEGB* are cyclic quadrilaterals and we have $\angle ADC = \angle ABE = \angle AGE$. Since $\angle FGA = 180° - \angle AGE = 180° - \angle ADC = 180° - \angle ADF$, quadrilateral *AGFD* is also cyclic. From the secants to the circumcircle of *AGFD*, we get $EA \cdot ED = EG \cdot EF$, and for the circumcircle of *AEGB*, we get $FB \cdot FA = FG \cdot FE$. By adding these two equations, we have $EA \cdot ED + FB \cdot FA = EG \cdot EF + FG \cdot EF = EF \cdot (EG + FG) = EF^2$.

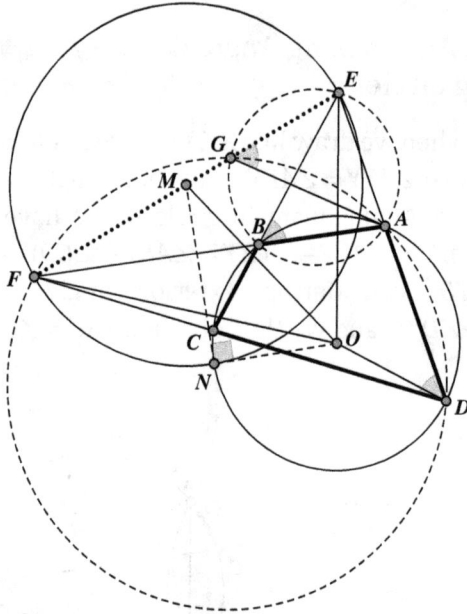

Figure 115-P

For the next step, we consider Figure 115a-P, where we again have the cyclic quadrilateral *ABCD*. As in Figure 115-P, point *E* is the intersection of line *DA* and *CB* and point *O* is the circumcenter of *ABCD*. We have also added a tangent *ET* of circle *O*, with point *T* as the tangency point. When from an external point a tangent and a secant are drawn to the same circle, the tangent is the mean proportional between the secant and its external segment, so that we have $\frac{ED}{ET} = \frac{ET}{EA}$, which we can also write as $EA \cdot ED = ET^2$. Since $EOT$ is a right triangle, we can apply the Pythagorean theorem to obtain $EA \cdot ED = ET^2 = OE^2 - OT^2$. Since $OD$ and $OT$ are both radii in circle $O$, we have $OT = OD$ and obtain $EA \cdot ED = OE^2 - OD^2$.

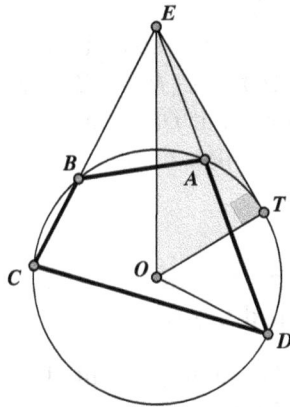

Figure 115a-P

Similarly, we also obtain $FB \cdot FA = OF^2 - OD^2$. Substituting these two expressions in $EA \cdot ED + FB \cdot FA = EF^2$ then gives us $OE^2 - OD^2 + OF^2 - OD^2 = EF^2$, which can also be written as $OE^2 + OF^2 + 2 \cdot OD^2 + EF^2 = 2 \cdot OD^2 + 4 \cdot ME^2$.

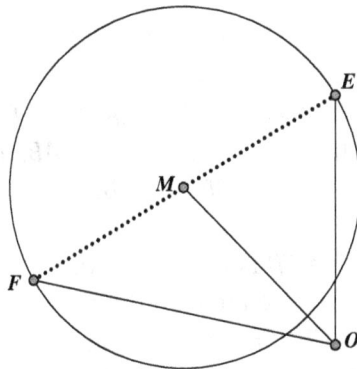

Figure 115b-P

Now, we are ready to consider Figure 115b-P, where we again have points $O$, $E$, $F$, and $M$ from Figure 115-P, with $ME = MF$, as $M$ is the center of the circle with diameter $EF$. Applying the law of cosines (see Toolbox) to triangle $OEM$ gives us $OE^2 = OM^2 + ME^2 - 2 \cdot OM \cdot ME \cdot \cos\angle OME$, and applying it to triangle $OMF$ gives us $OF^2 = OM^2 + MF^2 - 2 \cdot OM \cdot MF \cdot \cos(180° - \angle OME)$. Since we have $ME = MF$, adding these two equations gives us $OE^2 + OF^2 = 2 \cdot OM^2 + 2 \cdot ME^2$.

Recalling that we had established $OE^2 + OF^2 = 2 \cdot OD^2 + 4 \cdot ME^2$, we then obtain $2 \cdot OM^2 + 2 \cdot ME^2 = 2 \cdot OD^2 + 4 \cdot ME^2$, or $OM^2 = OD^2 + ME^2$. Since we know that $OD = ON$ and $ME = MN$, we can write this as $OM^2 = ON^2 + MN^2$ which establishes triangle $MNO$ as a right triangle, with $\angle ONM = 90°$.

## Curiosity 116. Circles That Generate an Equality of Seemingly Unrelated Segments

We begin Figure 116-P by drawing segments $GC$, $BC$, $CA$, and $BE$. Since angle $DBC$ is inscribed in a semicircle, we have $\angle DBC = 90°$, whereupon $\triangle GBC$ is a right triangle. The Pythagorean theorem gives us $CG^2 = BG^2 + CB^2$. Since $CG = CH$, as they are radii of circle $C$, we then have

$$CH^2 = BG^2 + CB^2. \tag{I}$$

We now note that $CH^2 = (CE + EH)^2 = CE^2 + EH^2 + 2CE \cdot EH = EH^2 + CE \cdot (CE + 2 \cdot EH) = EH^2 + CE \cdot CF$, which gives us

$$CH^2 = EH^2 + CE \cdot CF. \tag{II}$$

In triangle $BDF$, we have $\angle BFC = \angle ABC - \angle FCB$. Since point $C$ is the midpoint of arc $AB$, we have $\angle ABC = \angle CAB$, and since quadrilateral $ACEB$ is cyclic, we have $\angle BFC = \angle ABC - \angle FCB = \angle CAB - \angle FCB = (180° - \angle BEC) - \angle ECB = \angle CBE$.

This means that line $CB$ is tangent to the circumcircle of triangle $BEF$ because both angles are measured by one-half arc $BE$. From tangent $BC$ we have $CB^2 = CE \cdot CF$. Substituting this in equation (II) yields $CH^2 = EH^2 + CB^2$. When we combine this with equation (I), we find $BG = EH$, which is what we set out to prove.

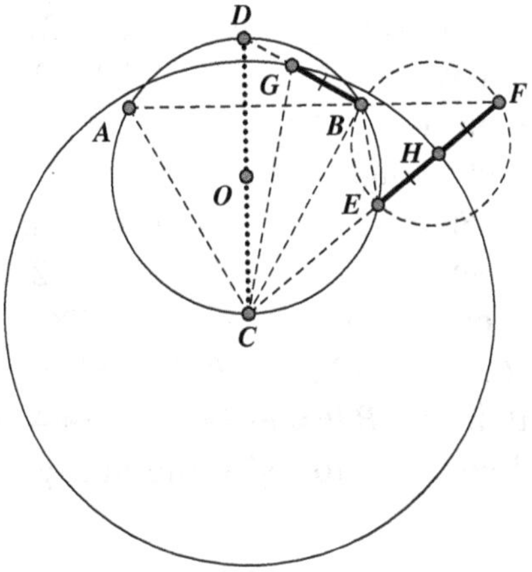

Figure 116-P

## Curiosity 117. Related Equilateral Triangles Inscribed in Concentric Circles

In Figure 117-P, we draw $B'O$ extended to intersect $A'B'$ at point $E$, and then add lines $OC'$, $OP'$, $PE$, and $PO$. Since triangle $A'B'C'$ is equilateral, line $B'OE$ is the perpendicular bisector of $A'C'$. In triangle $PA'C'$, we have $A'P^2 + C'P^2 = 2 \cdot PE^2 + 2 \cdot C'E^2$. (Recall that this was shown in

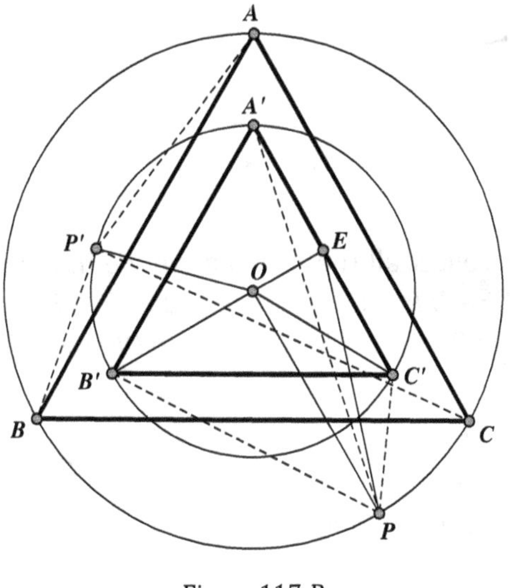

Figure 117-P

the proof of Curiosity 115 through Figure 115b-P.) Adding $B'P^2$ to both sides of this equation, we get $A'P^2 + B'P^2 + C'P^2 = 2 \cdot PE^2 + 2 \cdot C'E^2 + B'P^2$. Since $O$ is the common centroid of triangles $ABC$ and $A'B'C'$, we know that $B'E$ is a median of triangle $A'B'C'$, so that $OE = \frac{1}{2} \cdot OB'$.

We now consider Figure 117a-P, where applying the law of cosines to triangle $PEO$ results in $PE^2 = PO^2 + OE^2 - 2 \cdot PO \cdot OE \cdot \cos\angle POE$. Applying the law of cosines to triangle $POB'$ gives us $PB'^2 = PO^2 + OB'^2 - 2 \cdot PO \cdot OB' \cdot \cos(180° - \angle POE)$. Since $OB' = 2 \cdot OE$, this can also be written as $B'P^2 = PO^2 + 4 \cdot OE^2 + 4 \cdot PO \cdot OE \cdot \cos\angle POE$. Adding this second equation to twice the first yields $B'P^2 + 2 \cdot PE^2 = 3 \cdot PO^2 + 6 \cdot OE^2$.

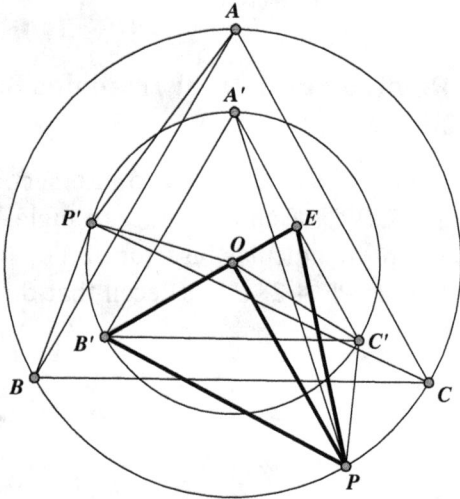

Figure 117a-P

We can now collect all the information we have thus far gathered. We know that $A'P^2 + B'P^2 + C'P^2 = 2 \cdot PE^2 + 2 \cdot C'E^2 + B'P^2 = (B'P^2 + 2 \cdot PE^2)$, and substitution of $B'P^2 + 2 \cdot PE^2 = 3 \cdot PO^2 + 6 \ OE^2$ then gives us

$$A'P^2 + B'P^2 + C'P^2 = 2 \cdot C'E^2 + (B'P^2 + 2 \cdot PE^2) = 2 \cdot C'E^2 + 3 \cdot PO^2 + 6 \cdot OE^2$$
$$= 2 \cdot (C'E^2 + OE^2) + 3 \cdot PO^2 + 4 \cdot OE^2$$
$$= 2 \cdot C'O^2 + 3 \cdot PO^2 + B'O^2$$
$$= 2 \cdot P'O^2 + 3 \cdot PO^2 + P'O^2$$
$$= 3 \cdot (PO^2 + P'O^2),$$

or $A'P^2 + B'P^2 + C'P^2 = 3 \cdot (PO^2 + P'O^2)$.

Similarly, we can also get $PA^2 + PB^2 + PC^2 = 3 \cdot (PO^2 + OP'^2)$, which proves the equality of the sums of squares.

## Curiosity 118. Intersecting Circles Surprisingly Generate Equal Lines

In Figure 118-P, we first draw the diameters $CBR$ and $DBS$ of the circle with center $B$. Since $\angle CER$ and $\angle DFS$ are each inscribed in a semicircle, they are both right angles. Furthermore, in circle $A$, the equality of angles subtended on arc $BP$ gives us $\angle RCE = \angle BCP = \angle BDP = \angle SDF$. This enables us to establish $\triangle ECR \cong \triangle FDS$, and thus, $CE = DF$.

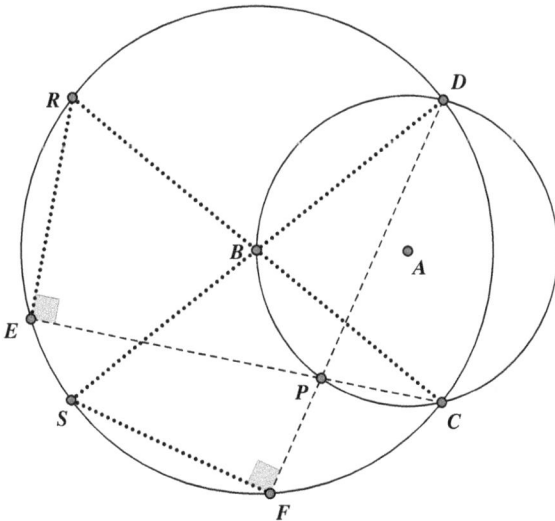

Figure 118-P

## Curiosity 119. Two Intersecting Circles Produce an Unexpected Line Segment Relationship

In Figure 119-P, we draw diameter *COD*, as well as lines *AO, BO,* and *AD*. Since *AQB* is a diameter of circle *Q*, triangle *AOB* is a right triangle. Similarly, since *COD* is a diameter of circle *O*, triangle *CAD* is also a right triangle. Because we have equal radii *OA* = *OC*, triangle *OCA* is isosceles, and we have $\angle OAC = \angle ACO$. Furthermore, in triangle *AOB*, we have $\angle ABO = 90° - \angle OAB = 90° - \angle OAC$, and similarly in triangle *CAD*, we have $\angle CDA = 90° - \angle ACD = 90° - \angle ACO$. Therefore, we have $\angle ABO = \angle CDA$, making $\triangle CAD \sim \triangle AOB$. From this, we obtain $\frac{AB}{DC} = \frac{AO}{AC}$, and since *DC* = 2*OC*, we then have $\frac{AB}{2OC} = \frac{OC}{AC}$, or $2OC^2 = AB \cdot AC$.

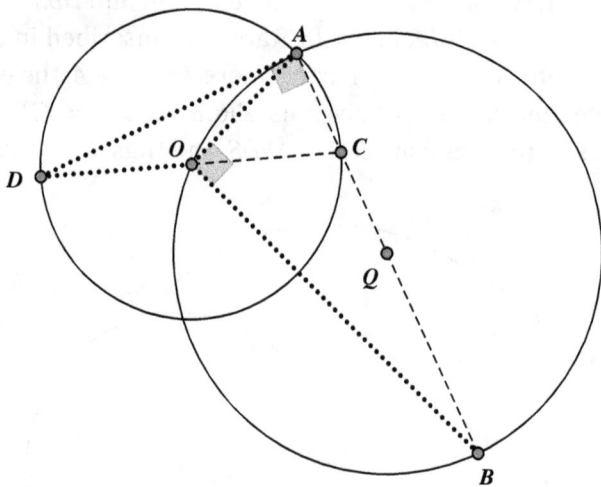

Figure 119-P

## Curiosity 120. A Quadrilateral with Two Intersecting Circles Generates Yet Another Circle

Since points $A$, $B$, $E$, and $F$ lie on the circumcircle of triangle $ABC$, as shown in Figure 120-P, we know that $\angle FEB = 180° - \angle BAF$. In triangle $BAF$, we have $180° - \angle BAF = \angle FBA + \angle AFB$, and we know that $\angle AFB = \angle ACB$, since they are both inscribed in arc $AB$. Furthermore, since points $A$, $P$, $Q$, and $C$ are concyclic, we have $\angle ACB = \angle ACQ = \angle BPQ = \angle BPR$, since both $\angle ACQ$ and $\angle BPQ$ are supplementary to $\angle QPA$. Thus, with $\angle FBA = \angle RBP$, we now have $\angle FEB = \angle FBA + \angle ACB = \angle RBP + \angle BPR = \angle FRP$. We notice that $\angle FRP$ is the exterior angle of triangle $BPR$. Since $\angle FEB = \angle FES$, we can write this as $\angle FES = \angle FRP$, from which we conclude that the points $E$, $F$, $R$, and $S$ are concyclic.

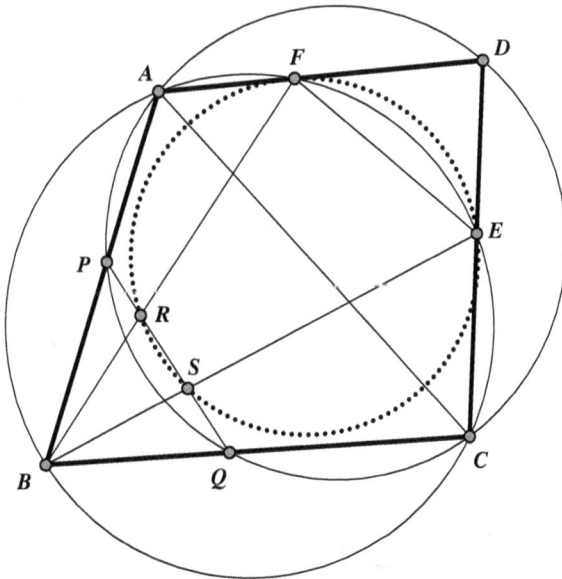

Figure 120-P

## Curiosity 121. The Broken Chord

To begin, we extend $BA$ to point $D$ so that $AD = AC$, as shown in Figure 121-P. This yields isosceles triangle $ACD$, so that $\angle DCA = \angle ADC$. We have $\angle BAC$ as an exterior angle of triangle $ACD$ and, therefore, $\angle BAC = \angle DCA + \angle ADC = 2\angle ADC$. In circle $O$, point $M$ is the upper intersection with the perpendicular bisector of $BC$. We notice that angle $BMC$ and angle $BAC$ are both measured by one-half arc $BC$, and we have $\angle BMC = \angle BAC = 2\angle ADC = 2\angle BDC$. We now consider the circle with center $M$ through points $B$ and $C$. Since $\angle BMC = 2\angle BDC$, point $D$ lies on this circle, and $AD$ is a chord. Since the perpendicular from the center of a circle to a chord bisects the chord, we have $BE = ED = AE + AD = AE + AC$, and thus, $BE = AE + AC$.

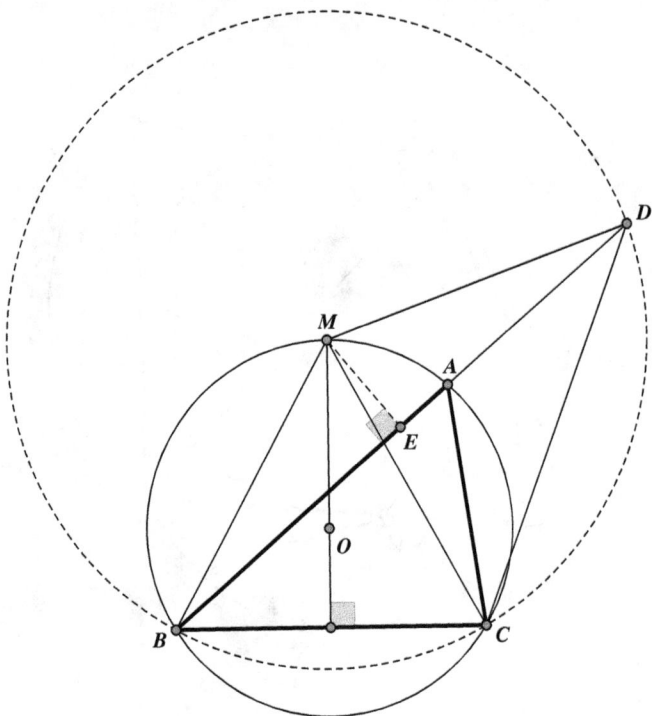

Figure 121-P

## Curiosity 122. Tangents Yield Equal Segments

In Figure 122-P, we begin by drawing lines $AB$ and $PD$. Our goal here is to prove that $\angle APD = \angle PDA$, which then implies that triangle $ADP$ is isosceles with $AD = AP$.

Since $AC$ is a tangent of circle $Q$, we have $\angle APD = \frac{1}{2}\overset{\frown}{ABD} = \angle CAD$, and since $AD$ is a tangent of circle $O$, we have $\angle BCA = \frac{1}{2}\overset{\frown}{AB} = \angle BAD$. We now note that $\angle CAD = \angle CAB + \angle BAD = \angle CAB + \angle BCA$, and since we have $\angle CAB + \angle BCA = \angle PBA$ in triangle $ACB$, and $\angle PBA = \angle PDA$ in circle $Q$, we can summarize to obtain $\angle APD = \angle CAD = \angle CAB + \angle BCA = \angle PBA = \angle PDA$. We have $\angle APD = \angle PDA$, and thus, $AD = AP$.

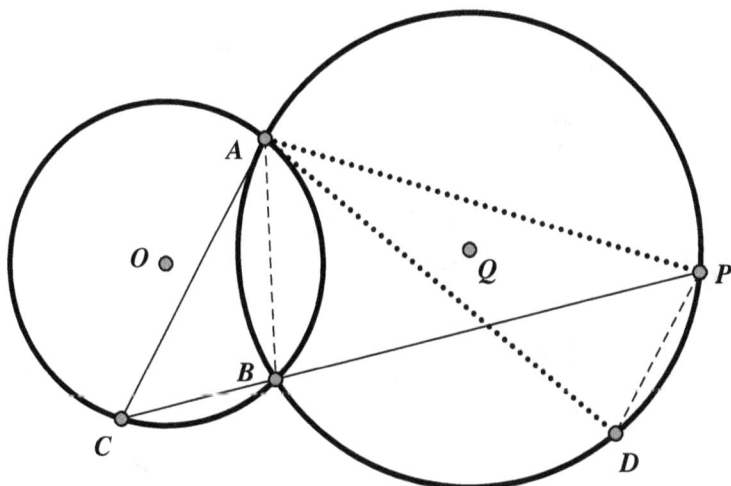

Figure 122-P

## Curiosity 123. The Fourth Corner of a Rectangle Lies on a Special Circle

In Figure 123-P, we draw the diameters $AX$ and $AY$ of circles $O$ and $Q$, respectively, as well as line $FY$. Since point $F$ lies on circle $Q$ with diameter $AY$, we have $\angle YFA = 90°$, and since $\angle AFE = 90° \angle AFE = 90°$ in rectangle $ADEF$, we see that points $E$, $F$, and $Y$ are collinear. Similarly, since point $D$ lies on circle $O$ with diameter $AX$, we have $\angle XDA = 90°$, and since $\angle EDA = 90°$ in rectangle $ADEF$, we see that points $D$, $X$, and $E$ are also collinear. We have therefore established that $\angle YEX = \angle BED$. Since $\angle BED = 90°$ in rectangle $ADEF$, we see that $\angle YEX = 90°$, and point $E$ must lie on the circle with diameter $XY$.

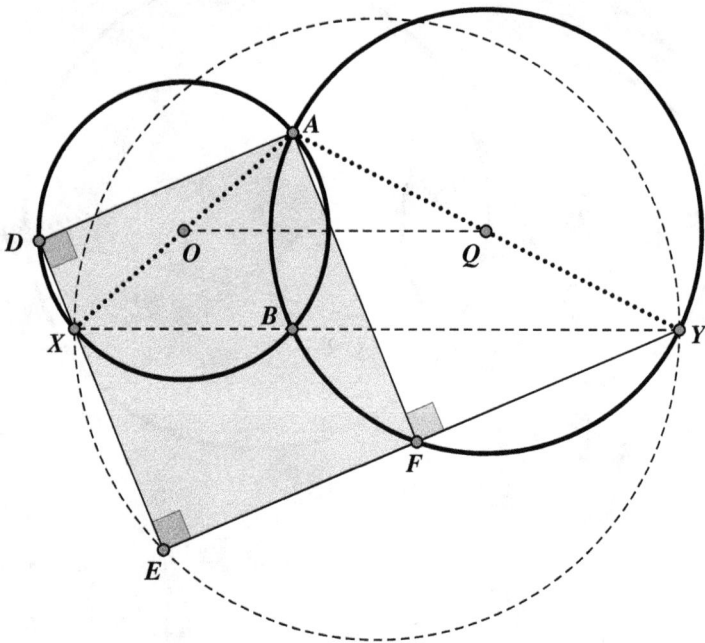

Figure 123-P

## Curiosity 124. Creating Equal Chords in Two Circles

In Figure 124-P, we draw the radii $OA$, $OC$, $QB$, $QD$, $MA$, and $MB$, as well as lines $QA$ and $OB$. In circle $M$, $AQ \perp AO$ and we have $AQ$ as a tangent of circle $O$ at point $A$. We also have $\angle AOC = 2 \cdot \angle QAB = \angle QMB$. We then have similar isosceles triangles $\triangle OAC \sim \triangle MQB$, and $\frac{AC}{AO} = \frac{BQ}{BM}$, or $AC = \frac{AO \cdot BQ}{BM}$. Similarly, we can also show that $\triangle QDB \sim \triangle MAO$, and this gives us $\frac{BD}{BQ} = \frac{AO}{AM}$, or $BD = \frac{AO \cdot BQ}{AM}$. Since $AM = BM = \frac{1}{2} \cdot OQ$, we thus have $AC = BD$.

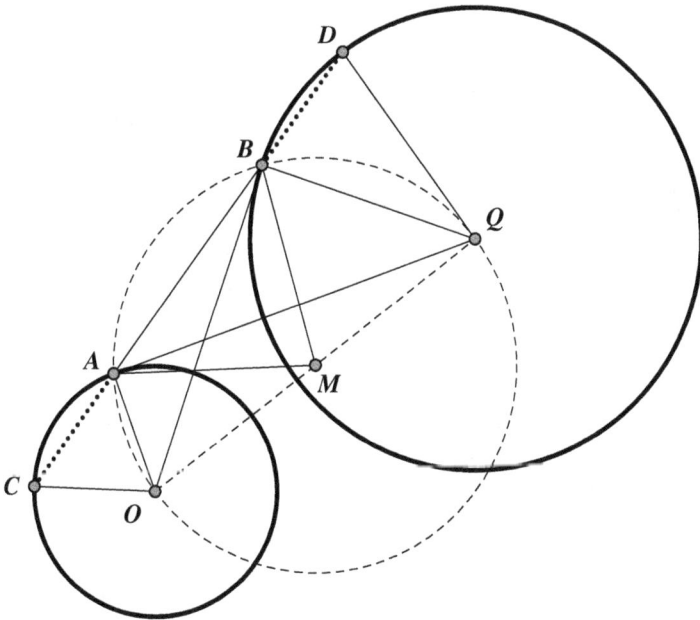

Figure 124-P

### Curiosity 125. An Unusual Angle Bisector Construction

As a first step, we draw lines *PA*, *PB*, *PD*, and *PE*, as shown in Figure 125-P. Since quadrilateral *APDC* is cyclic, we have $\angle PAE = \angle PAC = \angle PDB$, and since quadrilateral *BCEP* is also cyclic, we have $\angle AEP = \angle CBP = \angle DBP$. We are given $AE = BD$, and thus, we have $\triangle PEA \cong \triangle PBD$. It follows that $EP = PB$, and considering the angles in the circumcircle of quadrilateral *PBCE*, we then have $\angle ACP = \angle ECP = \angle PCB$. Thus, *CP* is the bisector of angle $\angle ACB$.

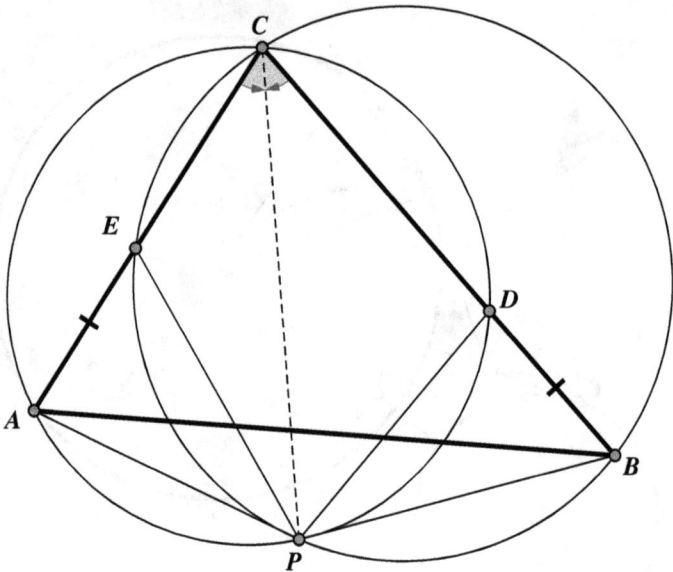

Figure 125-P

### Curiosity 126. A Square Derived from Two Circles with a Surprising Area

In Figure 126-P, since $MB = MC$ are radii in circle *Q*, and $OA = OC$ are radii in circle *O*, triangles *MBC* and *OAC* are isosceles. In triangle *MBC*, we have $\angle BMC = 180° - 2\angle CBM$, and in circle *M*, we have

$\angle BAC = \frac{1}{2}\angle BMC = 90° - \angle CBM$. From this, we obtain $\angle CAO = \angle BAO - \angle BAC = 90° - (90° - \angle CBM) = \angle CBM$, and therefore, $\triangle MBC \sim \triangle OAC$. Thus, we have $\frac{OA}{AC} = \frac{MB}{BC}$, or $\frac{OA}{MB} = \frac{AC}{BC}$. In an analogous way, we can also obtain $\triangle MCA \sim \triangle QCB$, and therefore, $\frac{MA}{AC} = \frac{QB}{BC}$, or $\frac{QB}{MA} = \frac{BC}{AC}$. We multiply these two equations to obtain $\frac{OA}{MB} \cdot \frac{QB}{MA} = \frac{AC}{BC} \cdot \frac{BC}{AC} = 1$. We can write $\frac{OA}{MB} \cdot \frac{QB}{MA}$ as $\frac{r}{MC} \cdot \frac{s}{MC} = \frac{r \cdot s}{MC^2} = 1$ so that $MC^2 = r \cdot s$. Since $MC^2$ is the area of the square with side $CM$, we find that its area is equal to the product $r \cdot s$.

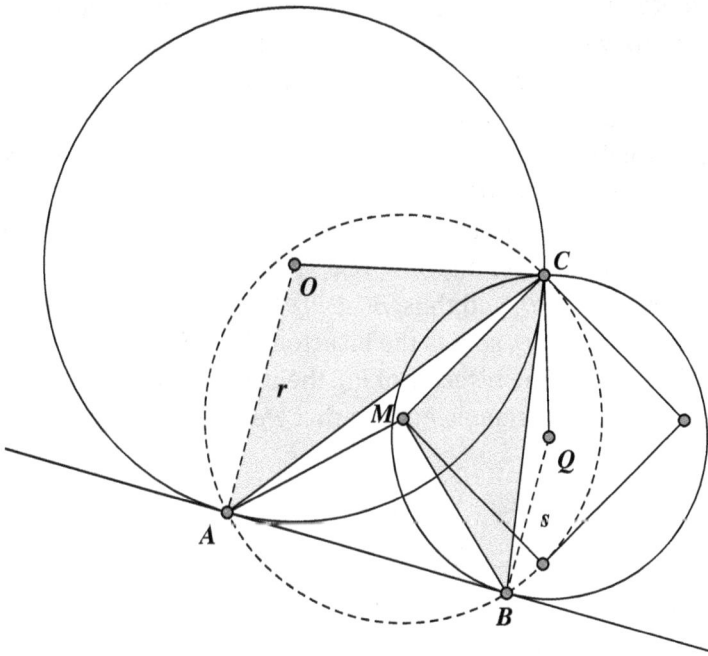

Figure 126-P

## Curiosity 127. An Unexpected Circumcircle Center

In Figure 127-P, we consider quadrilaterals *ABPD*, *ABRD*, and *ABQD*. The circle with center *B* and radius *AD* and the circle with center *D* and radius *AB* intersect in point *P* so that we have $AD = BP$ and $AB = DP$, and quadrilateral *ABPD* is a parallelogram. Since point *R* lies on side *BP* of parallelogram *ABRD* and quadrilateral *ABRD* is cyclic, we see that *ABRD* is an isosceles trapezoid with $BR||AD$ and $AB = DR$. This gives us $DP = AB = DR$, and then triangle *DRP* is isosceles with base *RP*. Since *AC* is a diameter of circle *O*, and point *D* lies on this circle, we have $DC \perp AD$. Since $BP||AD$, we also have $DC \perp BP$, and thus, $DC \perp RP$. We see that *DC* is an altitude in isosceles triangle *DRP*, and it is thus the bisector of *RP*. Similarly, since point *Q* lies on side *PD* of parallelogram *ABPD* and quadrilateral *ABQD* is cyclic, we see that *ABQD* is also an isosceles trapezoid with $DQ||AB$ and $AD = BQ$. This gives us $BQ = AD = BP$, and we see that triangle *BPQ* is isosceles with base *PQ*. Since *AC* is the diameter of circle *O*, and point *B* lies on this circle, we have $BC \perp AB$. Since $PD||AB$, we also have $BC \perp PD$, and thus, $BC \perp PQ$. We have *BC* as an altitude of isosceles triangle *BPQ*, so it is the bisector of *PQ*. Since *DC* is the bisector of *RP*, and *BC* is the bisector of *PQ*, their intersection point *C* is then the circumcenter of triangle *PQR*, so that we have $CP = CQ = CR$.

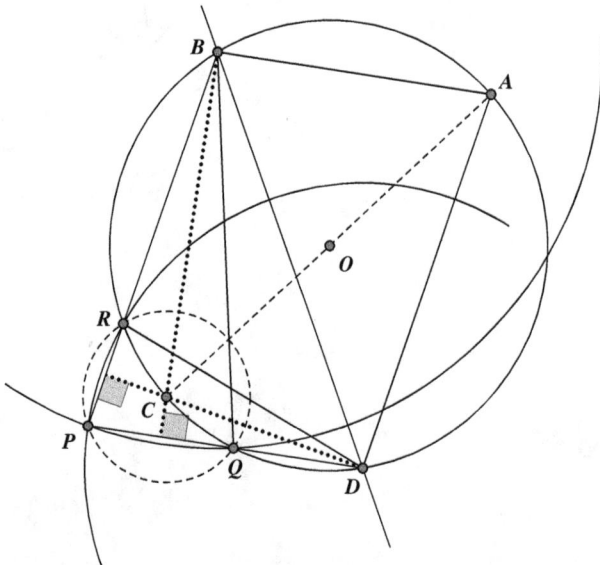

Figure 127-P

## Curiosity 128. Chords of a Constant Length

We begin by drawing lines $AB$, $AY$, and $BX$ in Figure 128-P. We also add an alternative point $P'$ along with the associated points $X'$ and $Y'$, as well as the lines connecting them with points $A$ and $B$. We note that $\angle PYA = \angle BYA = \angle BAX = \angle BXP$ in circle $O$, and that $\angle APY = \angle APB = \angle XPB$, which gives us $\triangle PAY \sim \triangle PXB$. From this, we obtain $\frac{PY}{PA} = \frac{PX}{PB} = a$, or $PY = a \cdot PA$, and $PY = a \cdot PB$, which implies $\triangle PAB \sim \triangle PXY$ and $XY = a \cdot AB$.

In order to prove that the length of line $XY$ is independent of the choice of point $P$ on circle $Q$, we will now show that $X'Y' = a \cdot AB$, with the same value of $a$. For this purpose, we consider triangles $\triangle BPX$ and $\triangle BP'X'$. Since $\angle BX'P' = \angle BX'A = \angle BXA = \angle BXP$ in circle $O$ and $\angle X'P'B = \angle AP'B = \angle APB = \angle XPB$ in circle $Q$, we see that $\triangle BPX \sim \triangle BP'X'$. This gives us $\frac{P'X'}{P'B} = \frac{PX}{PB} = a$, and thus, $P'X' = a \cdot P'B$. Analogously, for the similar triangles $PAB$ and $PX'Y'$, we obtain $P'Y' = a \cdot P'A$, and then $X'Y' = a \cdot AB$. We see that $XY = a \cdot AB = X'Y'$, so that the length of line $XY$ has a constant value, independent of the location of point $P$ on circle $Q$.

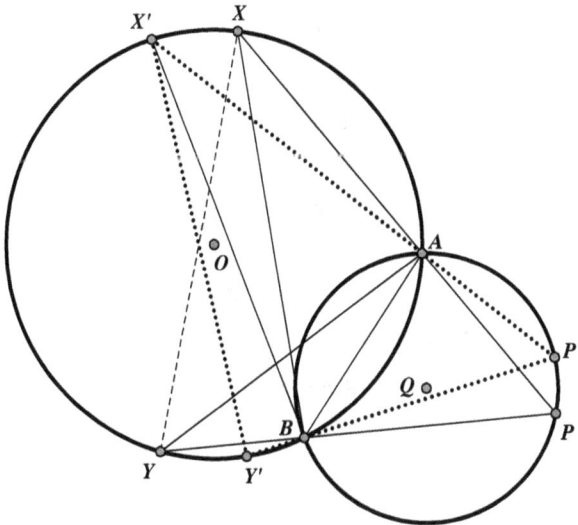

Figure 128-P

## Curiosity 129. An Unexpected Equality of Two Special Chords

In Figure 129-P, we begin by drawing lines $OQ$, $OB$, $AQ$, $BQ$, and $CQ$. Since the common chord $AB$ is perpendicular to the line $OQ$ joining the centers of the circles, we know that quadrilateral $OBQA$ is a deltoid (or kite), and triangle $QAB$ is isosceles. In circle $O$, we obtain $\angle AQB = 180° - \frac{1}{2}\angle BOA = 180° - \angle QOA$. Furthermore, we have $\angle BAC = \angle OAC - \angle OAB = 90° - \angle OAB = \angle QOA$. Since triangle $QAB$ is isosceles, we have $\angle BAQ = \frac{1}{2} \cdot (180° - \angle AQB) = \frac{1}{2}\angle QOA$, and we then have $\angle QAC = \angle BAC - \angle BAQ = \angle QOA - \frac{1}{2}\angle QOA = \frac{1}{2}\angle QOA$. Since triangle $QCA$ is also isosceles, both triangles $QAB$ and $QCA$ are isosceles with a common side $AQ$ and $\angle BAQ = \angle QAC = \frac{1}{2}\angle QOA$. Therefore, $\triangle QAB \cong \triangle QCA$ so that $AB = AC$.

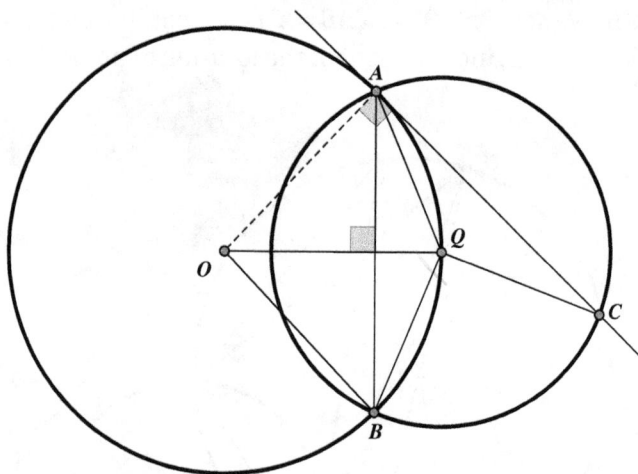

Figure 129-P

## Curiosity 130. A Surprising Common Point of Three Circles

In Figure 130-P, we draw triangle $OPQ$ and note that points $A$, $B$, and $C$ lie on its sides $PQ$, $QO$, and $OP$, respectively, since each pair of circles is symmetric with respect to the line joining their centers. We now consider the common tangent $YZ$ from $A$ to circles $P$ and $Q$, with $Y$ and $Z$ denoting the respective points of tangency. Since the radii $PY$ and $QZ$ of circles $P$ and $Q$ are perpendicular to this common tangent, triangles $AYP$ and $AZQ$ are both right triangles. With the equal vertical angles at point $A$, the triangles $AYP$ and $AZQ$ are similar. Letting $y = PY$ and $z = QZ$, which are the radii of circles $P$ and $Q$, respectively, we then get $\frac{AP}{AQ} = \frac{PY}{QZ} = \frac{y}{z}$. Letting $x$ denote the radius of circle $O$, in an analogous way for the other pairs of circles, we get $\frac{BQ}{BO} = \frac{z}{x}$ and $\frac{CO}{CP} = \frac{x}{y}$, and multiplying the three equalities gives us $\frac{AP}{AQ} \cdot \frac{BQ}{BO} \cdot \frac{CO}{CP} = \frac{y}{z} \cdot \frac{z}{x} \cdot \frac{x}{y} = 1$. From Ceva's theorem (see Toolbox), we can conclude that lines $OA$, $PB$, and $QC$ are concurrent at a point $E$.

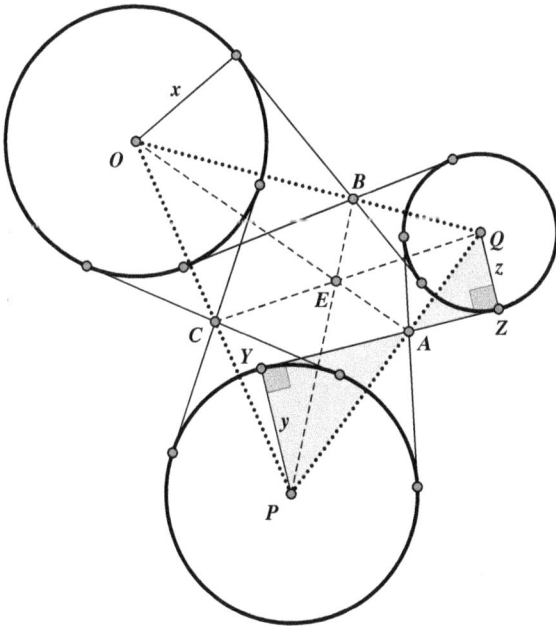

Figure 130-P

### Curiosity 131. Perpendicular Circles create a Diameter

In Figure 131-P, we first add lines $AQ$, $BQ$, $OQ$, $AB$, and $OB$. In circle $O$, since both angles are measured by arc $BA$, we have $\angle BOA = 2\angle BXA$. Since the line $OQ$ joining the circle centers bisects the angles $\angle BOA$ und $\angle AQB$, we have $\angle QOA = \angle BXA$. Since $\triangle AQO$ has a right angle in $A$, we get $\angle AQO = 90° - \angle QOA = 90° - \angle BXA$, and thus, $\angle AOB = 2\angle AQO = 2(90° - \angle BXA)$. In circle $Q$, we have $\angle BPA = 180° - \frac{1}{2} \cdot \angle AOB = 180° - (90° - \angle BXA) = 90° + \angle BXA$, and by triangle $BPX$ we get $\angle PBX = \angle BPA - \angle BXP = (90° + \angle BXA) - \angle BXA = 90°$. Since we now have $\angle YBX = \angle PBX = 90°$, $XY$ is a diameter of circle $O$.

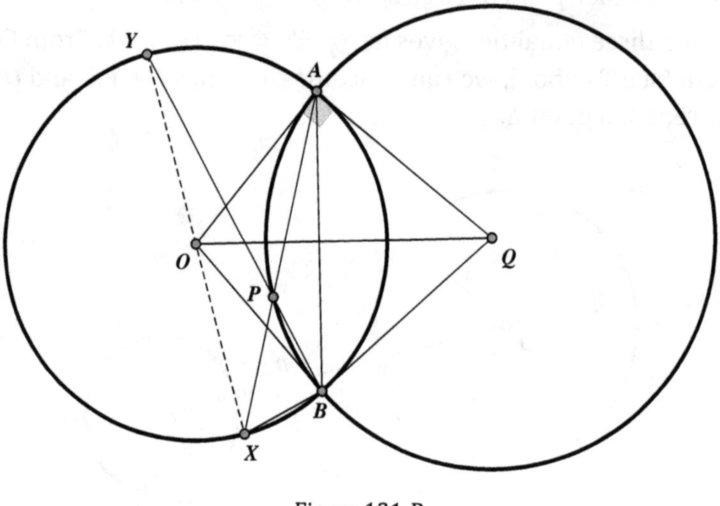

Figure 131-P

# Toolbox

# Introduction: The Geometry Toolbox

In the interest of providing the readership all the necessary "equipment" for the proofs in this book, and cognizant of the fact that some readers may have forgotten a few of the basics from the secondary geometry curriculum, we present this "Toolbox." This section contains some elementary concepts that aid appreciation of the amazing relationships which we hope you will have enjoyed experiencing in the first section of this book.

First, we will review some of the facts and principles you should be recognize from geometry class. These are presented without proof, as the reader's recollection should make them familiar once again. Naturally, any high school geometry book will furnish further understanding of these basic concepts.

Next, we present a number of more sophisticated theorems from basic trigonometry and somewhat advanced Euclidean geometry that may be familiar to some readers and less so to others, but are nevertheless easily understandable. The proofs accompanying these theorems demand only knowledge of the basic concepts from high school geometry. For the sake of brevity, the proofs are given for the most general cases, while special configurations may require some additional steps. Motivated readers may wish to explore some of these aspects.

The properties in the second part of the Toolbox are wonderful Quadrilateral Curiosities in their own right and could easily have been included in the core section of this book. They are considered "tools" because they are useful in more complex geometric proofs. Of course, this further enhances the particularly enchanting beauty of Euclidean geometry!

# A: Tools You Are Probably Familiar with from the High School Geometry Course

## A1: Congruence of Triangles

Triangles can be proved congruent by showing corresponding parts equal:

- Two sides and their included angle (SAS)
- Two angles and their included side (ASA)
- Two angles and a side not included (AAS)
- Three sides (SSS)
- For right triangles: the hypotenuse and a leg (HL)

## A2: Similarity of Triangles

Triangles can be proved similar by showing any of the following equivalent properties:

- The corresponding angles are equal. (AAA)
- The corresponding sides are proportional.
- Two pairs of corresponding sides are proportional and the included angles are equal.

## A3: Right Triangle Properties (see Figure A3)

For the following properties, we refer to a right triangle $ABC$ with its right angle at $C$. The lengths of the sides opposite vertices $A$, $B$, and $C$ are named $a$, $b$, and $c$, respectively. The altitude $CD$ on the hypotenuse $AB$ has length $h$ and divides the hypotenuse into segments $AD$ and $DB$ with lengths $p$ and $q$, respectively.

- The Pythagorean theorem: the sum of the squares of the legs equals the square of the hypotenuse: $a^2 + b^2 = c^2$.
- The square of the altitude $h$ on the hypotenuse is equal to the product of the lengths of the segments $p$ and $q$ on the hypotenuse: $h^2 = p \cdot q$.

- The square of the length $a$ of a leg is equal to the product of the length $c$ of the hypotenuse and the adjoining altitude segment $q$: $a^2 = c \cdot q$, also $b^2 = c \cdot p$.
- The median $CM$ to the hypotenuse of a right triangle is half the length $c$ of the hypotenuse $AB$.

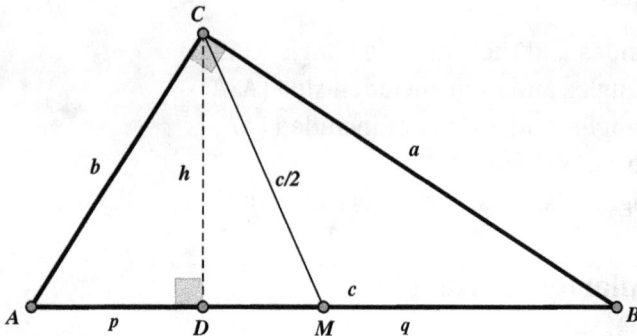

Figure A3

## A4: Angles Related to a Circle (see Figure A4)

These relationships are given in two different common notations, among the many in use internationally, in the hope that one of these will be sufficiently familiar for any reader.

- An inscribed angle: $\angle BAC = \frac{1}{2}\widehat{BC}$ (or $\angle CAB = \frac{1}{2}\cdot\angle COB$)
- An angle formed by 2 secants: $\angle F = \frac{1}{2}\left(\widehat{BC} - \widehat{DE}\right)$ (or $\angle CFB = \frac{1}{2}\cdot(\angle COB - \angle EOD)$)
- An angle formed by 2 tangents: $\angle J = \frac{1}{2}\left(\widehat{HAK} - \widehat{HCK}\right) = 180° - \widehat{HCK}$ (or $\angle HJK = \frac{1}{2}\cdot(180° - \angle KOH)$)
- An angle formed by a secant and a tangent: $\angle L = \frac{1}{2}(\widehat{CK} - \widehat{NK})$ (or $\angle KLC = \frac{1}{2}\cdot(\angle KOC - \angle NOK)$)

- An angle formed by 2 chords intersecting inside the circle: $\angle AME = \frac{1}{2}(\overset{\frown}{AE} + \overset{\frown}{BC})$ (or $\angle AME = \frac{1}{2} \cdot (\angle AOE + \angle COB)$)

- An angle formed by a chord and a tangent: $\angle JKC = \frac{1}{2}\overset{\frown}{KC} = \angle KAC$ (or $\angle JKC = \frac{1}{2}\angle KOC = \angle KAC$)

- Opposite angles in a circle (i.e., opposite angles of an inscribed quadrilateral): $\angle BCK = 180° - \angle KAB$

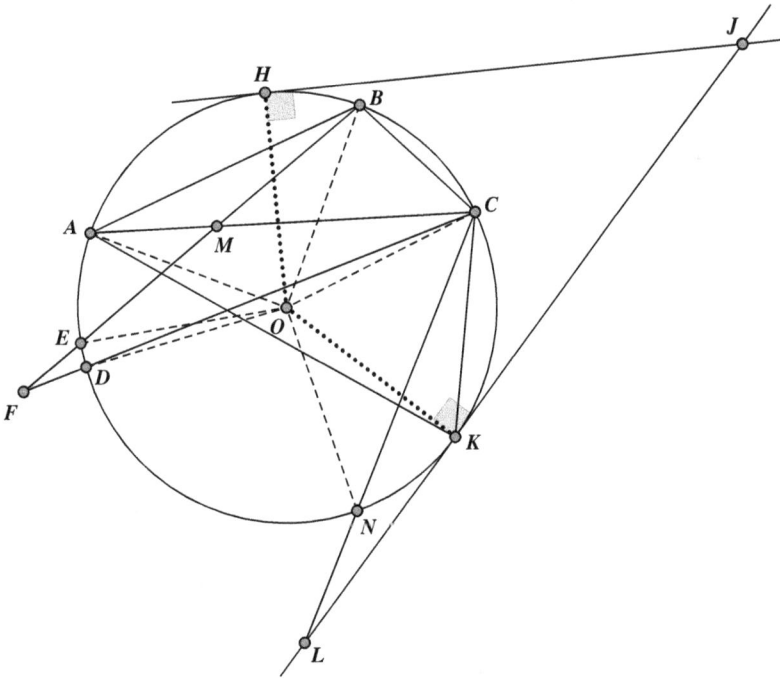

Figure A4

## A5:  Tangents, Secants, and Chords: Segments of a Circle and the Power of a Point (see Figure A5)

- From the same external point, the tangent is a mean proportional between the entire secant and its external segment: $\frac{e+f}{h} = \frac{h}{f}$ .
- From the same external point, for two secants: the product of the length of a secant and its external segment is equal to the product of the other secant and its external segment: $(e+f) \cdot f = (g+k) \cdot k$.
- For two chords intersecting inside a circle the product the segments of one chord is equal to the product or segments of the other chord: $a \cdot b = c \cdot d$.

The products $(e+f) \cdot f = (g+k) \cdot k$ and $a \cdot b = c \cdot d$ are referred to as the *power* of the point in which the secants and chords intersect, respectively.

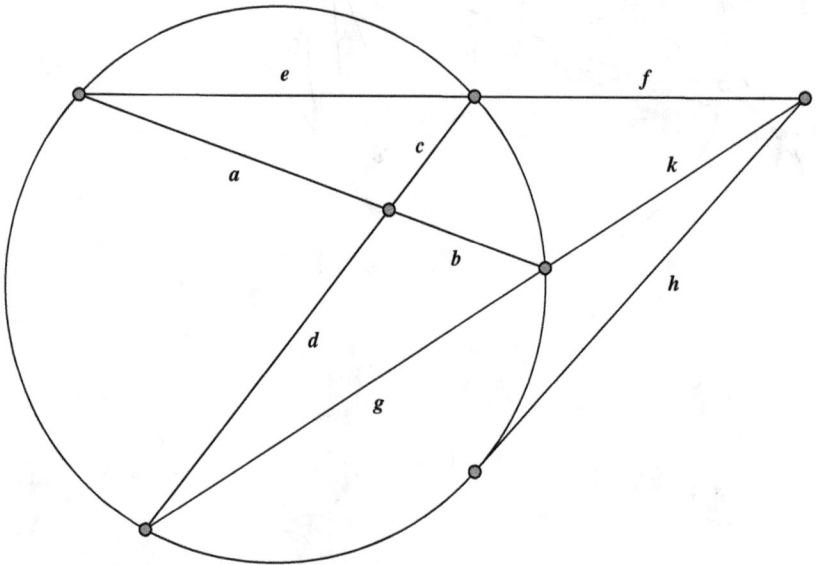

Figure A5

## A6: The Law of Sines and the Law of Cosines (see Figure A6)

In triangle $ABC$, we let $a$, $b$, and $c$ denote the lengths of the sides opposite the vertices $A$, $B$, and $C$, respectively.

Then $\frac{a}{\sin \angle A} = \frac{b}{\sin \angle B} = \frac{c}{\sin \angle C}$. This is known as the *Law of Sines*.

Also, $c^2 = a^2 + b^2 - 2ab \cdot \cos\angle ACB$, and analogously, $a^2 = b^2 + c^2 - 2bc \cdot \cos\angle BAC$ and $b^2 = c^2 + a^2 - 2ca \cdot \cos\angle CBA$. This is known as the *Law of Cosines*.

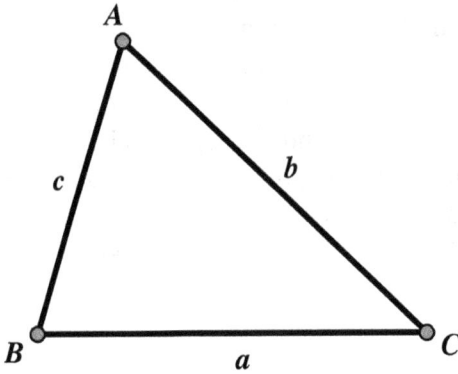

Figure A6

# B: Less Familiar Tools – However, Useful and Fascinating

## B1: Ceva's Theorem

A common topic in triangle geometry concerns lines joining the vertices of a triangle with points on their opposite sides. Such lines are often called *cevians*, named after the Italian mathematician Giovanni Ceva (1647–1734), who published his famous theorem in 1678. The result is very powerful and leads to many geometric wonders. Ceva's theorem states that three cevians are concurrent if and only if the product of the lengths of the alternate segments made by the points of contact on the sides are equal. We see this illustrated in Figure B1a, where triangle $ABC$ has cevians $AD$, $BE$, and $CF$, meeting the sides $BC$, $CA$, and $AB$ at points $D$, $E$, and $F$, respectively. In the figure, they are shown to meet at a common point $P$. According to Ceva's theorem, the cevians $BC$, $CA$, and $AB$ intersect in a common point $P$, if and only if $AF \cdot BD \cdot CE = AE \cdot BF \cdot CD$, or, equivalently, $\frac{AE}{CE} \cdot \frac{BF}{AF} \cdot \frac{CD}{BD} = 1$.

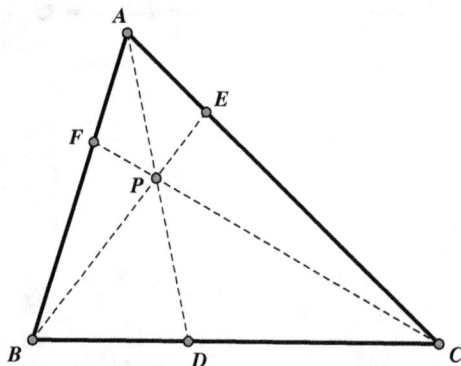

Figure B1a

**Proof:** We will divide this proof into steps. First, we assume that the three cevians $BC$, $CA$, and $AB$ intersect at a common point $P$. We will show that $\frac{AE}{CE} \cdot \frac{BF}{AF} \cdot \frac{CD}{BD} = 1$. In order to show this, we consider Figure B1b.

Here, starting from the configuration in Figure B1a, we have added a line parallel to $BC$ through $A$. This line intersects $BE$ extended at point $R$ and $CF$ extended at point $S$.

The parallel lines enable us to establish several pairs of similar triangles and relationships between the lengths of their sides resulting from these similarities. We have:

$$\Delta AER \sim \Delta CEB, \text{ and therefore, } \frac{AE}{CE} = \frac{AR}{BC}, \tag{I}$$

$$\Delta BCF \sim \Delta ASF, \text{ and therefore, } \frac{BF}{AF} = \frac{BC}{AS}, \tag{II}$$

$$\Delta CPD \sim \Delta SPA, \text{ and therefore, } \frac{CD}{AS} = \frac{DP}{AP}, \tag{III}$$

$$\Delta BDP \sim \Delta RAP, \text{ and therefore, } \frac{BD}{AR} = \frac{DP}{AP}. \tag{IV}$$

From (III) and (IV) we get $\dfrac{CD}{AS} = \dfrac{BD}{AR}$, or $\dfrac{CD}{BD} = \dfrac{AS}{AR}$. $\tag{V}$

Now we multiply (I), (II), and (V) to obtain our desired result: $\frac{AE}{CE} \cdot \frac{BF}{AF} \cdot \frac{CD}{BD} = \frac{AR}{BC} \cdot \frac{BC}{AS} \cdot \frac{AS}{AR} = 1$.

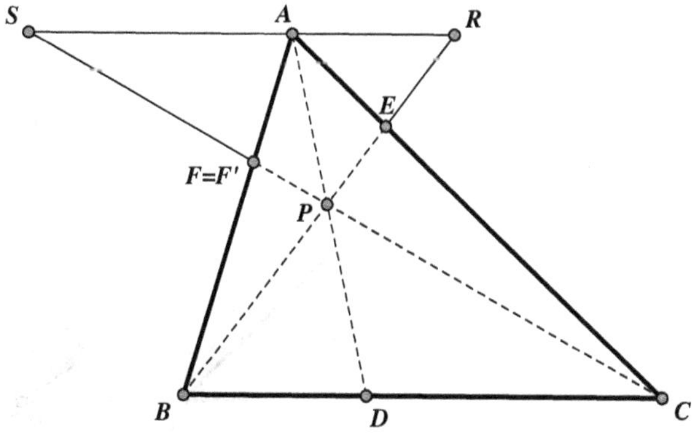

Figure B1b

Next, we assume that points $D$, $E$, and $F$ are given on the sides $BC$, $CA$, and $AB$, respectively, of triangle $ABC$, with $\frac{AE}{CE} \cdot \frac{BF}{AF} \cdot \frac{CD}{BD} = 1$. We wish to show that line segments $AD$, $BE$, and $CF$ then have a common point $P$. Suppose $AD$ and $BE$ intersect at $P$. Draw $CP$ and call its intersection with $AB$ point $F'$. Since $AD$, $BE$, and $CF'$ are concurrent, we can use the part of Ceva's theorem we have already proved to state $\frac{AE}{CE} \cdot \frac{BF'}{AF'} \cdot \frac{CD}{BD} = 1$. Since our hypothesis stated that $\frac{AE}{CE} \cdot \frac{BF}{AF} \cdot \frac{CD}{BD} = 1$, we obtain $\frac{BF'}{AF'} = \frac{BF}{AF}$, and thus, F = F', proving the concurrency.

## B2: Ceva's Theorem Extended

Ceva's theorem also holds true when the cevians intersect outside the triangle. In Figure B2a, the cevians $AD$, $BE$, and $CF$ meet the sides $BC$, $CA$, and $AB$ at points $D$, $E$, and $F$, respectively. Here, the configuration is such that they intersect outside the triangle. In this case, it is also true that $AD$, $BE$, and $CF$ meet in a common point $P$ if and only if,
$$\frac{AE}{CE} \cdot \frac{BF}{AF} \cdot \frac{CD}{BD} = 1.$$

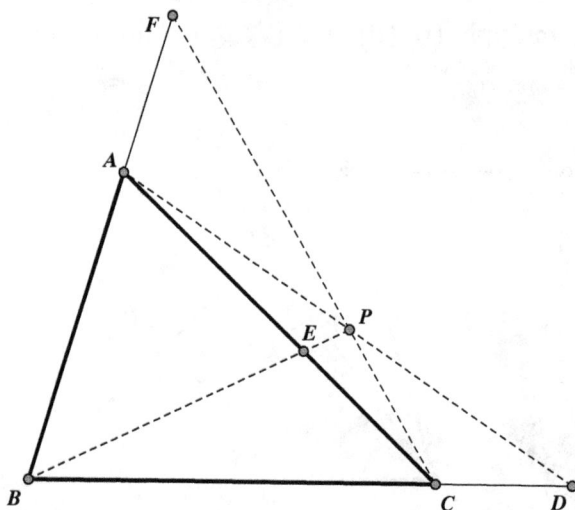

Figure B2a

**Proof:** The proof in this case is identical to the proof when point $P$ is in the interior of $ABC$. In Figure B2b, we have added points $R$ and $S$ in the same way as in Figure B1b, and the proof from B1 is identical for this configuration.

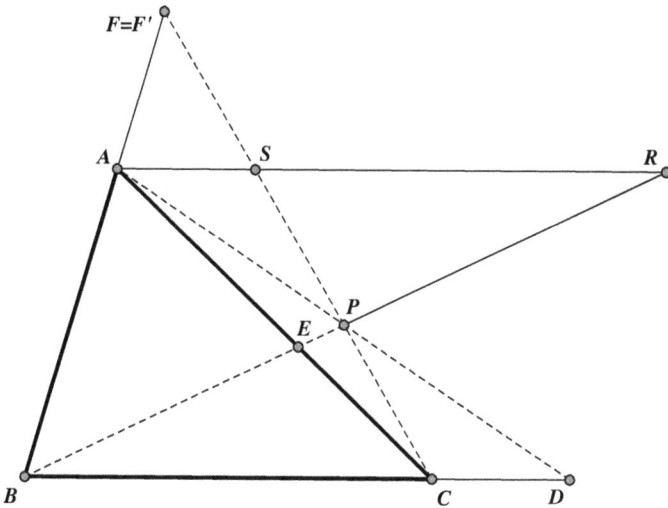

Figure B2b

## B3: Ptolemy's Theorem

The product of the lengths of the diagonals of a cyclic quadrilateral equals the sum of the products of the lengths of the pairs of opposite sides. In other words, in Figure B3, we have $AC \cdot BD = AB \cdot CD + BC \cdot DA$.

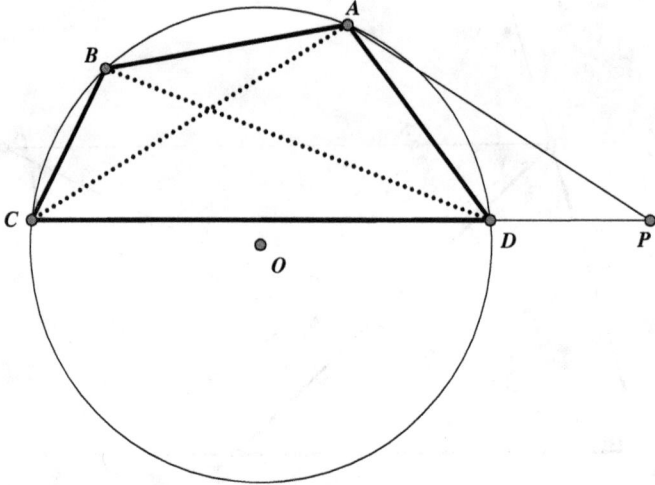

Figure B3

**Proof:** In Figure B3, quadrilateral $ABCD$ is inscribed in circle $O$. A line is drawn through $A$ to meet $CD$ extended at $P$, so that $\angle BAC = \angle DAP$. Since quadrilateral $ABCD$ is cyclic, $\angle CBA$ is supplementary to $\angle ADC$. Since $\angle PDA$ is also supplementary to $\angle ADC$, we have $\angle CBA = \angle PDA$, and because of the way point $P$ was constructed, triangles $BCA$ and $DPA$ are similar. We then have $\frac{AB}{DA} = \frac{BC}{DP}$, or $DP = \frac{BC \cdot DA}{AB}$. Furthermore, we also get $\frac{AB}{DA} = \frac{AC}{AP}$. Since $\angle BAD = \angle BAC + \angle CAD = \angle DAP + \angle CAD = \angle CAP$, triangles $ABD$ and $ACP$ are also similar. We then also have $\frac{BD}{CP} = \frac{AB}{AC}$, or $CP = \frac{AC \cdot BD}{AB}$. We know that $CP = CD + DP$, and substituting for $CP$ and $DP$, we get $\frac{AC \cdot BD}{AB} = CD + \frac{BC \cdot DA}{AB}$, or $AC \cdot BD = AB \cdot CD + BC \cdot DA$.

## B4: Isometries: Reflection and Rotation

In the course of a geometric proof, it is often useful to define a point transformation that maps a configuration of points onto a congruent configuration. Such transformations are known as *isometries*. The defining characteristic of an isometry is that it maps every point in the plane onto a unique point in such a way that the distance between any two points is equal to the distance between their images. This means that any geometric figure will always be mapped onto a congruent figure by the isometry.

Two important examples of isometries are *reflections* and *rotations*. These are illustrated in the following figures. In Figure B4a, we see a reflection defined by line *l*. We say that a point *P* is *reflected* about line *l* onto a point *P'*, where *P'* is the point *symmetric* to *P* with respect to *l*. From the point *P*, a line perpendicular to line *l* through point *P* intersects line *l* at point *L*. Point *P'* is then defined as the point on this line with $LP = LP'$ and $P \neq P'$. (Note that if a point *P* lies on line *l*, we define $P = P'$.)

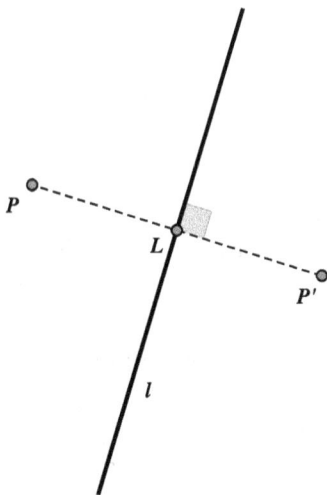

Figure B4a

In Figure B4b, we see a rotation defined by center $O$ and angle $\theta$. We say that a point $P$ is *rotated* about $O$ through $\theta$ onto a point $P'$. From the point $P$, line segment $OP$ is drawn. Point $P'$ is then defined as the point with $\angle POP' = \theta$ and $OP = OP'$. (Note that, for the sake of completeness, we define that $O$ is rotated onto itself, independent of the size of angle $\theta$.)

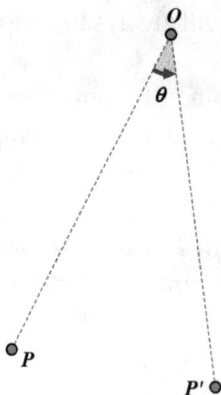

Figure B4b

In order to prove that these transformations are indeed isometries, we must show that all line segments are transformed onto line segments of equal length.

In Figure B4c, we see a line segment $PQ$ reflected about line $l$ onto a line segment $P'Q'$. In order to prove that $PQ = P'Q'$, we draw lines $MP$ and $MP'$, where $M$ is the intersection of line $l$ and $QQ'$. Since $LP = LP'$ and $PP' \perp LM = l$, triangles $LPM$ and $LMP'$ are congruent, and we have $MP = MP'$ and $\angle LMP = \angle P'ML$. This allows us to note $MQ = MQ'$, $\angle PMQ = \angle LMQ - \angle LMP = \angle Q'ML - \angle P'ML = \angle Q'MP'$, and $MP = MP'$, which show us that triangles $MPQ$ and $MQ'P'$ are also congruent. From this, we see that $PQ = P'Q'$ holds.

Note that we assumed here that $P$ and $Q$ both lie on the same side of $l$. If they lie on opposite sides of line $l$, or one or both lie on line $l$, the proof must be modified a bit, but this is left to the intrepid reader as an easy exercise.

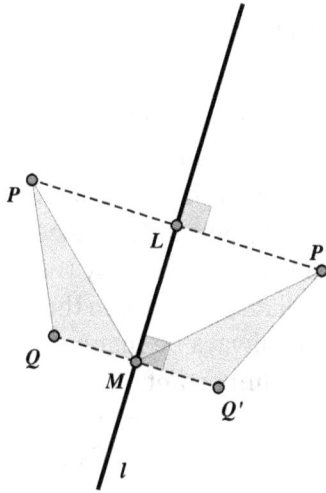

Figure B4c

Similarly, in Figure B4d, we see a line segment $PQ$ rotated about $O$ through $\theta$ onto a line segment $P'Q'$. In order to prove that $PQ = P'Q'$, we consider triangles $OPQ$ and $OP'Q'$. Since $OP = OP'$, $OQ = OQ'$, and $\angle POQ = \angle POP' - \angle QOP' = \theta - \angle QOP' = \angle QOQ' - \angle QOP' = \angle P'OQ'$, we see that triangles $OPQ$ and $OP'Q'$ are congruent. From this, we again see that $PQ = P'Q'$. As above, we note that the details of this proof depend on the relative positions of the points, but fully analogous proofs will work for any other positions, and these are left to the intrepid reader.

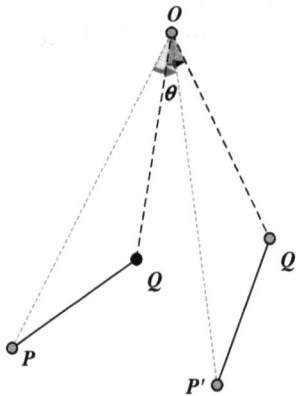

Figure B4d

## B5: Homothety and Similarity

A slightly more general point transformation that is also frequently useful maps a configuration of points onto a geometrically similar configuration. Such transformations are known as *similarity transformations*. The defining characteristic of a similarity transformation is the fact that it maps every point in the plane onto a unique point in such a way that the distance between any two points, multiplied by a constant, is equal to the distance between their maps. This means that any geometric figure will always be mapped onto a similar figure by such a transformation. An important example of a similarity transformation is *homothety*. This is illustrated in Figure B5a.

A *homothety* is a mapping of points $P$ in a plane onto corresponding points $P'$ for which there exists a fixed point $O$ and a fixed factor $a$ (that is, a real number $a \neq 0$) such that $SP' = a \cdot SP$ for all points $P$.

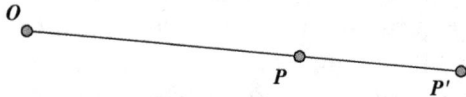

Figure B5a

The most important property of a homothety is that it maps any line segment $PQ$ onto a line segment $P'Q'$ with $PQ||P'Q'$ and $P'Q' = a \cdot PQ$. In order to prove this, we consider Figure B5b. Since $OP' = a \cdot OP$ and $OQ' = a \cdot OQ$, and triangles $OPQ$ and $OP'Q'$ share a common angle in $O$, they are similar. We therefore have $a = \frac{OP'}{OP} = \frac{OQ'}{OQ} = \frac{P'Q'}{PQ}$, whereupon it follows that $P'Q' = a \cdot PQ$. Furthermore, since $\angle QPO = \angle Q'P'O$, we also have $PQ||P'Q'$.

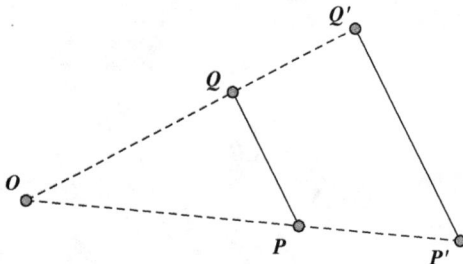

Figure B5b

# Index